MAN IN THE MANGROVES

The Socio-economic Situation of
Human Settlements in Mangrove Forests

Proceedings of a workshop held at
Nong Nuch Village, Pattaya, Thailand, 27–31 May 1985,
sponsored by the United Nations University and
the National Research Council of Thailand

**Edited by Peter Kunstadter, Eric C. F. Bird,
and Sanga Sabhasri**

THE UNITED NATIONS UNIVERSITY

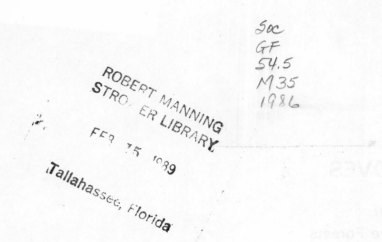

More than half the world's people live in coastal regions, utilizing such resources as salt, minerals, fish, and crustaceans, the products of mangroves, salt marsh, seagrass, and kelp, energy from wind, waves, and tides, and such materials as sand, gravel, clay, and limestone, all obtained from the coast or the adjacent sea. Moreover, the coast provides sites for settlement, agriculture and aquaculture, ports and harbours, industry, commerce, and recreation. The management of coastal environments and their resources has raised many problems in both developed and developing countries, and it was felt appropriate that the United Nations University should give emphasis to this field of study.

The Coastal Resources Management Project was initiated as part of the University's Natural Resources Programme. It was decided that the coastal environment — comprising the foreshore (between high and low tide lines), backshore (above high tide line to the landward limit of marine influences), and nearshore (from low tide line out to a depth of 20 metres) zones — was a distinctive field for research and training that merited its own project within the programme.

A number of research studies and workshops were commissioned under this theme. *Man in the Mangroves* contains papers presented at a UNU-sponsored workshop. Three of the papers result from UNU research. The remainder were submitted by independent researchers. They focus on the socio-economic aspects of the use, development, and management of mangrove areas in relation to environmental and ecological factors.

Although the Coastal Resources Management Project has now been concluded, the University's new programme on Resource Policy and Management has undertaken to maintain an international dimension in research, training, and dissemination, stressing the interaction of resource management, conservation, and development.

The United Nations University
Toho Seimei Building, 15-1 Shibuya 2-chome, Shibuya-ku, Tokyo 150, Japan
Tel.: (03) 499-2811   Telex: J25442   Cable: UNATUNIV TOKYO

Printed in Japan

NRTS-29/UNUP-607
ISBN 92-808-0607-6
United Nations Sales No. E.86.III.A.7
01500 P

CONTENTS

CONTENTS

PREFACE

The papers which follow are edited versions of presentations given at the Workshop on the Socio-economic Situation of Human Settlements in Mangrove Forests, held at Nong Nuch Village, Pattaya, Thailand, 27–31 May 1985, under the auspices of the United Nations University and the National Research Council of Thailand.

Following the introductory addresses, the results of studies on socio-economic aspects of the use, development, and management of mangrove areas were considered in relation to ecological and environmental factors. A general introduction (Kunstadter) is followed by case studies of the situation in Thailand (Aksornkoae et al., and Puckprink), Malaysia (Chan), Indonesia (Mantra), Australia (Bird), Sri Lanka (Silva), Tanzania (Mainoya et al.), Japan (Miyawaki), and South America (Snedaker). The papers are followed by a set of recommendations based on a review of the situation in Thailand by the workshop participants in the perspective of the other case studies.

These case studies fall short of a global review (notable omissions include the Philippines, Papua New Guinea, New Zealand and other Pacific island nations, Burma, Bangladesh, India, Pakistan, the Arabian peninsula, parts of Africa, and Central America). They nevertheless represent a reasonably comprehensive range of environmental, ecological, and socio-economic variations in human–mangrove relationships and the associated problems of development.

Many generalizations emerged from the diverse national settings described in the papers and from the ensuing discussion. Many of these are mentioned in the introductory chapter, and others are included in recommendations, which, although they refer specifically to Thailand, may have wider applicability.

The workshop was part of the activity of the United Nations University within a wider programme area on Resource Policy and Management (originally initiated in 1976 under the designation "Use and Management of Natural Resources"). One important bottleneck impeding the development of the right resource-use systems in the tropics is the need for further research. Thus the UNU Resource Policy and Management programme has focused on a number of issues in projects such as Assessment Studies on Arid Lands Management (1977–1984), Water–Land Interactive Systems (1977–1984), Highland–Lowland Interactive Systems (since 1977), Agro-forestry Systems (since 1977), and Coastal Resources Management (1977–1985) (United Nations University 1982). The objectives of these projects are connected with an international discussion on the interaction between population, resources, environment, and development, known by the acronym PRED.

As an example of the PRED interaction, one could mention the well-known phenomenon of shifting cultivation, or land-rotation with bush fallow. This land-use system can be regarded as stable if the fallow periods are long enough for the regeneration of soil and vegetation. If, however, the pressure of population increases and the fallow intervals are shortened, the vegetation growth is reduced and soils are degraded. The system loses its productivity, and the subsistence of the population is no longer ensured. Previous solutions to this problem have been to introduce irrigated agriculture or tree crops, both of which have been successful in South-East Asia. However, a vision of the tropical zone as an exporter of tree crops and an importer of food is not realistic, particularly as the demand for tropical products is rather stagnant and land resources are becoming increasingly scarce. Also, even if countries earn a considerable amount of foreign exchange through exports of cash crops, they often need these earnings for other essential imports (e.g. in the field of technology) and not for food alone (Waller 1984). These considerations are part of the rationale for a land- and resource-use system such as agro-forestry, which combines the production of food and wood (including firewood) and at the same time is useful as a tool for management of resources and conservation of the environment. It can be considered a nearly closed system, because it requires few

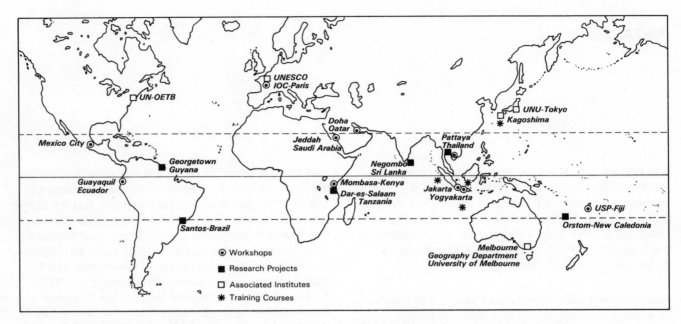

Global network of the UNU Coastal Resources Management project, 1978–1985

costly inputs such as chemical fertilizers and pesticides.

Because of the food crises, especially now in tropical Africa, further research, training, and dissemination of information on system requiring low external input are very important. One of the greatest weaknesses in new and appropriate forms of resource-use systems such as agroforestry, eco-farming, and aquaculture is the lack of management packages that can be implemented under specific conditions of climate, soil, and vegetation (Ruddle and Manshard 1981). This weakness is also characteristic of the mangrove ecosystem in the tropical and subtropical world.

The UNU work on mangroves is part of the Coastal Resources Management project, (co-ordinated by Eric Bird, of the University of Melbourne). Several studies on the use and management of coastal and near-shore resources in tropical environments were initiated under this project. A number of graduate research and training courses were sponsored in Indonesia (Jakarta, Yogyakarta, Sunda Strait) by the UNU. Several UNU fellowships to universities and research institutions were awarded, notably in Japan and the United States, and a series of international workshops in Fiji, Jeddah, Mombasa, and Paris was organized. Work started in 1983 on the traditional uses and socio-economic implications of eco-system changes in tropical mangrove areas. Case studies were prepared in Thailand, Tanzania, and

Sri Lanka, results of which are discussed in papers in this volume. The locations of activities and institutions associated with this project are shown on the map above.

Work on coastal resources is also linked to the newly established UNU project on Climatic, Biotic, and Human Interactions in the Humid Tropics, which was started in 1983. While this project concentrates mainly on bio-physical and geo-physiological problems (e.g. the effects of deforestation and land use on soil, hydrology, microclimate, and productivity), the human side is reflected in seven case studies on the resource uses of pioneer settlements in frontier zones of the humid tropics.

In sum, it is felt that there is not nearly enough applied and pure research on the very complex development thresholds of tropical countries. A combined effort of many institutions and organizations is needed for improved policy implementation. It is in this sense that this workshop is a contribution, as mandated in the Charter of the United Nations University, to solving some of "the pressing global problems of human survival, development and welfare."

— Walther Manshard
Programme Director
Development Studies Division
The United Nations University

References

Bird, E. C. F. 1984. ''United Nations University — Coastal Resources Project.'' United Nations University, Tokyo.

Ruddle, K., and W. Manshard. 1981. *Renewable natural resources and the environment.* Tycooly International, Dublin.

United Nations University. 1982. *The Natural Resources Programme: 1977–1981.* United Nations University, Tokyo.

Waller, P. 1984. ''The ecological handicaps of the tropics.'' *Intereconomics* (Hamburg), 3: 137–142.

WELCOME AND OPENING ADDRESS

WELCOME

Mr. Sombhan Panateuk
Director, Sriracha Regional Forest Office, Royal Forest Department

Professor Sanga Sabhasri, Professor Walther Manshard, and distinguished guests, on behalf of the Eastern Regional Forest Office here in Sriracha, I would like to welcome all of you from overseas and from Thailand.

I am very pleased that the United Nations University and the National Research Council of Thailand have arranged this Workshop on the Socioeconomic Situation of Human Settlements in Mangrove Forests, at Pattaya in our region.

We have about 40,000 hectares of mangrove forest growing in estuaries in this region. The mangrove forest has been used for many years, mostly for charcoal production. We now face problems with people who live in and near the mangrove area. They still cut the mangrove for firewood and charcoal production on a sustained-yield basis, but now some of them have cleared the forest for aquaculture, especially for shrimp farms. This raises questions about the possibility of sustained use of the mangrove forest resources.

I have learned from Dr. Sanit Aksornkoae that excellent mangrove ecologists, both from overseas and from Thailand, have come to this workshop. The knowledge obtained from this workshop will be very useful to us, and also to our friends from neighbouring countries, for mangrove resource management and development. I hope that the workshop will be very fruitful and that all of you will enjoy your stay.

OPENING ADDRESS

Professor Sanga Sabhasri
Permanent Secretary, Ministry of Science, Technology, and Energy, Government of Thailand

It is a great honour for me to have the opportunity to participate in the Workshop on the Socioeconomic Situation of Human Settlements in Mangrove Forests, which is jointly organized by the United Nations University and the National Research Council of Thailand.

When we speak about mangrove forests, everyone now agrees that these forests provide us with important natural resources extremely beneficial to people living along the coastal areas and nearby. Mangrove forests make up only about 15.8 million hectares, or 0.6 per cent of all inland forests in the world. About 6.5 million hectares, or 41.4 per cent of the world total, are found in tropical Asia. Although small in comparison with the world's total forests, they play a very important role in the ecosystem of the region. They prevent soil erosion by acting as a wind and water break. They maintain moisture and breeding grounds for many plants and animals both on land and in the sea. They also provide food, construction materials, fibres, and medicinal plants to dwellers in and near the coastal zones.

Problems of exploitation of mangrove resources are increasing due to the rapid recent growth of the population. In the late 1960s the complex pressures resulting from population growth, urban expansion, and economic development brought about heavy exploitation and destruction of mangrove resources. Detrimental activities included poorly executed logging operations, alluvial mining, road construction, and conversion of mangrove forests into shrimp farms, fish ponds, and salt pans. In addition, many mangrove forests near big cities have been reclaimed for real estate developments. It has recently been noted that many mangrove areas have been manipulated beyond their environmental tolerance. The over-exploitation of mangrove resources, without concern for their maintenance, reflects the outmoded view that mangroves are an inexhaustible resource. The time has come to realize that such an attitude towards the mangrove environment needs to be changed, and a sense of responsibility to protect the mangroves must be restored.

Public awareness of these problems began in the early 1970s. There were several incidents which first drew the attention of Thai scientists. Among them were the effects of military use of herbicides on mangroves in South Viet Nam. This attracted the attention of American, European, and Vietnamese scientists to questions of productivity and regeneration of mangroves. Thai scientists were invited by the US National Academy of Sciences to join in the study of these problems. Mangrove areas in Thailand were used as a baseline for study of the ecosystem of relatively undisturbed mangroves. Plants in undisturbed mangrove forests were investigated to determine their ecological role, as a comparison with mangroves destroyed by military use of herbicides [*The effects of herbicides in South Vietnam,* National Academy of Sciences, Washington, D.C., 1974].

Coinciding with the study of effects of herbicides was the conversion of mangroves to fish ponds at an alarming rate. This drew attention from Thai conservationists and scientists. A group of Thai scientists who were deeply concerned with the loss of mangrove forests met in the early 1970s. They produced a message to the public that over-exploitation and misuse of the mangrove ecosystem could lead to detrimental effects on economically important aquatic species along the coastline such as fish, prawns, shrimp, crabs, and oysters. An appeal was issued stating that a symbiotic relationship in which humans are equal partners with nature must be recognized, mangrove resources had to be renewed, and a better management system had to be developed quickly to restore stability to the mangrove ecosystem. To achieve these objectives, the scientists asked that basic information should be collected and understood and that this information should be used as a basis for improved management and utilization. In order to mobilize resources for this purpose, support was sought from international organizations such as Unesco and the United Nations University in addition to bilateral organizations.

Problems in the management and conservation of mangrove resources can be classified into three key needs. The first is to re-establish a stable mangrove ecosystem after the exploitation of the forest area; the second, to maintain the relationship between forest and fisheries; and the third, to enhance the function of mangrove forests in erosion control. Both information on human uses and scientific knowledge of the natural ecosystem are required in order to develop effective management and conservation of the mangrove ecosystem. Education and training for public appreciation of the mangrove system must be increased, and awareness and consciousness of the importance of the mangrove ecosystem must be established among high-level decision-makers.

As the mangrove area is losing ground to fish ponds, shrimp farms, and other uses, it is recommended that the carrying capacity of the mangrove area should be determined before converting it to other resource development projects. The socio-economic aspects of human activities in mangrove areas should be taken into account. The socio-economic consequences of a decision by the state to allow a few private entrepreneurs to take control of the area should be taken into account. In general, the mangrove resources in South-East Asia are owned and managed by the state, and it is considered that the general public has an interest in this property, but mangrove dwellers depend specifically on mangrove resources for their livelihood. The decision to give ownership of concessions to a few entrepreneurs would eventually cause great hardship for mangrove dwellers, most of whom are poor. Therefore, a balanced relationship between fishery production and forest production in mangrove areas is necessary in order to benefit the largest number of people. It is essential that countries which possess these valuable coastal resources should concentrate their political will and aim their highest policies at sustained yield from the mangrove resources, while moving toward greater equity and a more even distribution of the income and other benefits from these resources among rural people.

Successful policy planning for the development and management of mangrove resources depends on many factors. Policy planning and implementation cannot be successful without basic data on dwellers in the mangrove forest and on the dynamics of the watershed areas and the coastline.

This workshop is a great occasion for experts and policy-makers from different corners of the world to have an opportunity to meet and share their views and experience. I believe that the four days of the workshop will yield pertinent knowledge which can be applied to successful policy planning for the management of mangrove forests. I believe this workshop, attended by distinguished researchers, will produce appropriate recommendations for policy planning to develop, manage, and maintain mangrove resources. I am sure that, as long as we are aware of the significance of mangrove resources, mangrove forests will continue to exist and will be preserved as useful natural resources not only for all of us today but also for future generations.

Finally, I want to take this opportunity to express my appreciation to the United Nations University, to the National Research Council of Thailand, and to all those who made this workshop possible, and to welcome all the participants.

1. SOCIO-ECONOMIC AND DEMOGRAPHIC ASPECTS OF MANGROVE SETTLEMENTS

Peter Kunstadter

Introduction

This paper reviews some of the progress made over the past 15 years in studies of socio-economic and demographic aspects of human use of mangrove areas, with emphasis on South-East Asia. Mangroves are usually defined in terms of the distribution of characteristic tree species. Mangrove forests are found growing in brackish water on the margin between land and sea in tropical and subtropical areas, but, as with other definitions of ecological zones, this is not completely satisfactory because the exchange of individual organisms, nutrients, and energy between this margin and both the sea and upstream areas is at least as important as what goes on within the geographic limits of mangrove tree species (fig. 1).

Discussions of the socio-economic aspects of human settlements in the mangroves are difficult for several reasons. Mangrove forest ecosystems and the socio-economic systems of mangrove settlers are not coterminous, and, as compared with the natural ecosystem, information on the socio-economic systems of mangrove dwellers is sparse. Most of it deals with economics (e.g. yields of forestry and fisheries), and relatively little describes the social and human ecological systems of the human residents.

Long-term residents of mangrove areas are generally similar ethnically to the inland populations (see, e.g., Aksornkoae et al. 1984, 34ff.), but their way of life often involves adaptation to mangrove environmental conditions, and economic exploitation of several distinct ecological zones. Thus mangrove dwellers have many different socio-economic systems, some of which are primarily focused on subsistence activities (including both agriculture and fishing) and some are primarily commercial (including agriculture, fishing, and forestry).

Traditional mangrove dwellers often combine the use of land, sea, and inter-tidal resources. Even with limited economic development and moderni-

zation, the boundaries of the social and economic systems that influence mangroves spread well beyond the ecological limits of the zone itself. For example, charcoal from mangroves has long been an item of international trade, as have fish and shellfish. With increasing economic development, the boundaries of the socio-economic systems that use mangrove resources spread even further. The future fate of the mangroves (e.g. with respect to extractive forest harvest or the hazards of pollution from oil transported by sea, or the spread of industrial development to and along the coastline, or the development of seaside resorts or condominiums) may be decided in the air-conditioned board rooms of temperate-zone businesses. Thus the proper study of the socio-economics of mangroves must include some attention to national and even international social, economic, and demographic processes (fig. 2).

Although rich in many resources, mangrove forests have traditionally been sparsely settled and unintensively exploited by humans. In South-East Asia this has probably been the result of the scarcity of fresh water for domestic use and the unsuitability of mangrove soils for long-term agricultural exploitation; in Papua New Guinea it may have been the consequence of the presence of highly effective mosquito vectors of malaria. Modernization (e.g. development of motorized transportation or the extension of modern urban water supplies) has allowed the spread of human populations into mangrove areas without requiring that the newcomers adapt to mangrove ecological conditions. At the same time, modern technology has made the clearing of mangroves and their conversion to other uses such as shrimp ponds or urban dwelling sites relatively inexpensive, thereby opening mangrove areas for more intensive exploitation, with rapid environmental change and rapid expansion of the human population.

Because forests are one of the major mangrove resources, studies of mangroves have often been conducted as forestry research. Fifteen years ago the emphasis in studies of all aspects of forestry

1

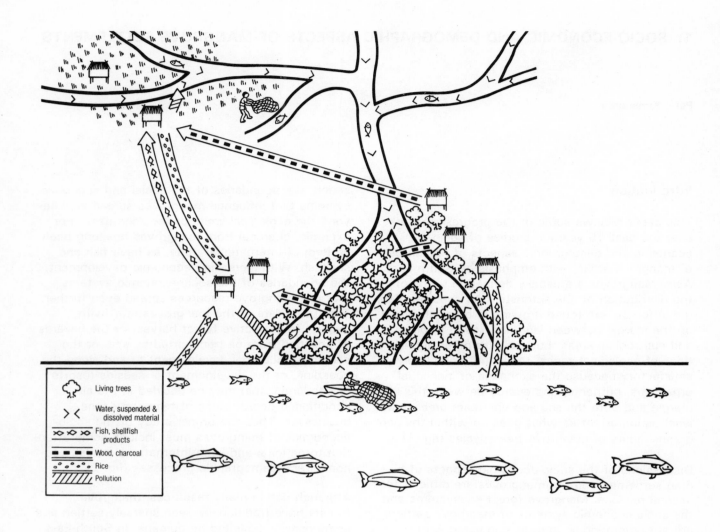

FIG. 1.   Characteristics of mangrove management systems: traditional subsistence economies

was on the upland forests, while mangroves were "the forgotten forests." Following the massive destruction of mangroves associated with the military use of herbicides in Viet Nam, and the attention focused on those forests by the American Association for the Advancement of Science and the National Academy of Sciences (e.g. Committee on the Effects of Herbicides in Vietnam 1974, sec. IV C), scientists from Thailand were drawn into studies of their own, relatively well preserved mangroves. Largely as a result of the efforts of Sabhasri (1979) and Aksornkoae (e.g. Aksornkoae et al. 1984), mangrove forests in Thailand are now among the best studied and best known in the world. As attention was being drawn to mangrove forests, the importance of mangrove fisheries and spawning grounds was also being recognized (see, e.g., MacNae 1974; Martosubroto and Naamin 1977).

Several ecological and economic characteristics of mangroves are now relatively well understood:
— Mangrove forests perform multiple ecological functions (e.g. production of woody trees; provision of habitat, food, and spawning grounds for fish and shellfish; provision of habitat for birds and other valuable fauna; protection of coastlines and accretion of sediment to form new land), and some of these functions have benefits far beyond the geographical limits of the mangrove zone itself (Hamilton and Snedaker 1984; MacNae 1974; Saenger, Hegerl, and Davie 1983).
— Mangrove areas have high biological productivity associated with heavy leaf production and leaf fall, and rapid decomposition of the detritus (see, e.g., Christensen 1978; Knox and Miyabara 1984, app. 1).
— The mangrove ecosystem is dynamic, changing

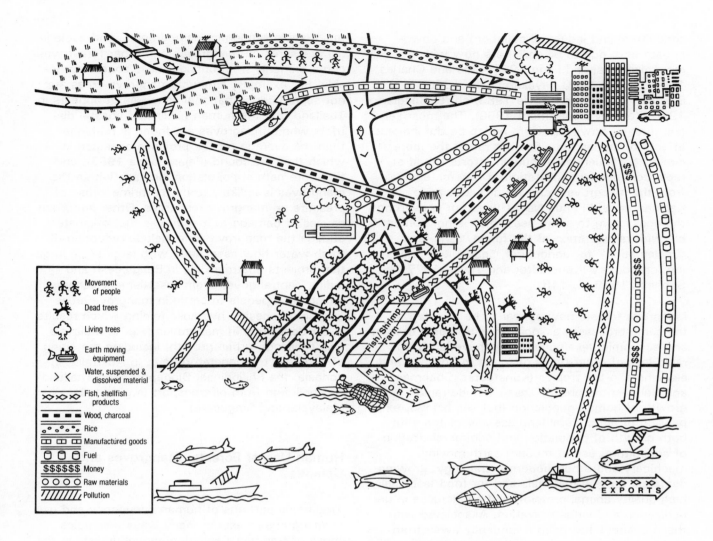

FIG. 2.   Characteristics of mangrove management systems: modern market economies

in both location and composition, and has great resilience with the ability to restore itself after heavy damage, as long as seed sources and water flow are maintained (Committee on the Effects of Herbicides in Vietnam 1984; Lai and Feng 1984).

— There are many direct economic benefits from mangrove resources (mangroves may be, e.g., a source of firewood and charcoal, self-renewing sites for collecting fish and shellfish, sites for collecting honey, an attraction for tourism) (Hamilton and Snedaker 1984, chs. 3, 5, 6, 9).

— In comparison with irrigated farms and upland forests, the relatively undisturbed mangrove forests in South-East Asia provide a salubrious environment for people, associated with the absence of vectors for important diseases (Kunstadter, in press). In the relatively undisturbed state, mangrove forests do not support the

reproduction of the local brackish-water malarial mosquito *Anopheles sundaicus* (Arwati 1977; Boschi 1975, 1976; Desowitz et al. 1974). This is in contrast with New Guinea, where highly effective malaria vectors (*A. farauti*) breed in undisturbed mangrove areas (Smithurst 1970; Spencer 1964, 1976), so that malaria inhibits human settlement in these zones (Petr 1977).

— Mangrove zones are useful for a great variety of human settlements, ranging from villages for near-subsistence fisherfolk to housing and industrial developments (Hamilton and Snedaker 1984, ch. 10).

Despite their natural resilience, mangroves are threatened on a worldwide scale with unprecedented widespread and long-term damage or destruction. The harmful consequences of human activities upstream (e.g. due to various forms of agricultural and industrial pollution) and down-stream (e.g. due to oil spills) are now widely recognized (Lai and Feng 1984; Chia and Charles 1984; Gomez 1980; Piyakarnchana 1980; Saenger, Hegerl, and Davie 1983, ch. 5; Soegiarto 1980; Zoology Department 1980). The most serious threats, however, seem not to be the indirect effects of human activities but rather the more direct consequences of human efforts aimed at rapid commercial exploitation of raw materials from the mangroves (e.g. for cellulose as in Kalimantan and Papua New Guinea) or converting the ecosystem to some use which is not compatible with regeneration of the forest, as in land "reclamation" for agriculture, aquaculture, housing, or industry (Kunstadter and Tiwari 1977, 3; Sabhasri 1979, 5, 14ff.).

Mangrove forests now appear to be affected by the same processes associated with modernization and economic development that have led to the rapid loss of other types of forest in Thailand and elsewhere in the tropics (Kunstadter, Chapman, and Sabhasri, in press). These include rapid growth of human population (but see below), expansion of agricultural land-use associated with both growth of population and commercialization of agriculture, use of modern earth-moving machinery and other modern technology, growing demands for raw materials and for food (especially high-quality animal protein), and an increase in urbanization and industrialization. Associated with these changes has been a tendency away from traditional patterns of sustained multiple use and toward increased specialization of use of land in ways that, for at least the short run, reduce the options for (or greatly increase the costs of) other uses (e.g., once a mangrove forest is converted to salt farms, the area is generally unsuitable for agriculture or regeneration to forest).

Mangroves, like highland forests, are generally considered "marginal," not in the sense of being unproductive but in the sense of being relatively remote and quite different from cities and farms. Even the terminology applied to mangrove forests (at least in English) tends to be derogatory (mangroves are generally termed "swamps" rather than forests or fish hatcheries), whereas the discussion of mangrove destruction ("reclamation") tends to be euphemistic, implying or stating that the mangroves are wasted in their uncleared or undrained state.

The implications of research on population change, as well as of research on pollution, are that the major socio-economic, demographic, and ecological pressures on mangroves and other forest types come primarily from outside these zones, not inside. For example, although mangrove areas are not segregated in national censuses, it appears in Thailand that the human fertility of coastal districts where mangroves are found is no higher than the average for the province or region in which they are found (Pejaranonda 1985), and therefore natural population increase within the mangroves is unlikely to be the prime cause of pressure on mangrove resources. Other important causes of damage to mangroves that originate outside the mangrove zone include reduction of fresh-water flow associated with large-scale irrigation projects upstream (as in Bangladesh and India), central-government resettlement schemes which send people to settle in mangrove areas (e.g. in Indonesia), hydraulic mining which results in silt deposition in mangroves (e.g. in Malaysia and Thailand), urban growth, including the siting of large electric generators (e.g. in Australia, Indonesia, the Philippines, Singapore, and Thailand), and pollution from oil transport (e.g. in Indonesia, Malaysia, and Singapore).

Human Use of Forests: Mangroves and Uplands

Traditional patterns of human occupancy and use of mangrove forests in many ways resembles those of traditional swidden agriculturalists in the hills. Population density was low and settlements were relatively small. Technology was based on simple materials, minimal use of fossil fuels, and small numbers of tools. The use of the technology (e.g. in terms of fishing gear) was often ingenious and reflected an intimate knowledge of the environment. The traditional economy was largely self-contained. Long-term settlers in general had developed a conservative, self-sustaining adaptaion to the local environment. Studies of contemporary mangrove communites (e.g. Aksornkoae et al. 1984) suggest the continuing importance of subsistence production and home consumption of local products even in mangrove communities that are at least partially integrated into the market economy.

Social controls were based on a moral community, that is, consensus of the villagers regarding desirable goals, the appropriate ways of achieving them, and sanctions against those who violated norms. These norms tended to perpetuate the long-term conservation of the environment within which the people expected their descendants to live a life similar to the one they experienced.

Traditional use systems in the highlands as well as in the mangroves involved clear-cutting of patches. Under more conservative systems that did not destroy seed stock for regeneration and involved short periods of use followed by long fallow, there was natural regeneration of the forest when it was abandoned (Kunstadter, Chapman, and Sabhasri 1978; Saenger, Hegerl, and Davie 1983, chs. 4 and 5). Similar patterns were followed in mangrove forests that were carefully managed for long-term commercial exploitation, e.g. in Malaysia.

In the highlands wood from the cut forest was used as fuel to convert above-ground biomass to fertilizer for agriculture; in the mangroves, wood cut from the forest was used and often sold as fuel or as building material, and soil fertility was restored by infusions from upstream. Under conditions of low population density and a subsistence-oriented economy, both systems provided long-term sustained yield. Both systems generally involved multiple use of the varied forest environment at least for subsistence (e.g. hunting or fishing for food, collecting for minor food resources, building materials, and medicines).

In both systems the existence of a transitional zone between forested and non-forested areas is associated with increased species diversity, to the benefit of the human residents. In the highlands this is accomplished by successional forest regrowth after the swidden is abandoned or fallowed (Kunstadter 1979). The mangrove forest is itself a transitional zone between dry land and open ocean, which contains a richly varied environment.

Several socio-economic and ecological problems associated with economic development are common to highlanders and sea-margin dwellers. These include failure to protect the traditional inhabitants' interest in the land, damage to the environment as a result of the physical processes of development, harvest at rates far higher than natural replacement, emphasis on single uses that restrict employment rather than on multiple uses, and, in general, loss of traditional life-style.

Traditional patterns of land ownership and land use are often unrecognized by the authorities. The government claims the land as ''forest'' or

''wasteland'' which may be turned over to new owners or concessionaires for commercial exploitation or settlement. Although this may provide wage-labour opportunities for traditional inhabitants, commercial use as forest (which emphasizes short-run economic efficiency and profit) may not support as many people as were supported by traditional land-use systems. Aksornkoae et al. (1984, 110–111), for example, show that the number of people employed in mangrove concessions is only about 17 per cent of those who actually live in mangrove areas. Moreover, the beneficiaries from jobs created by the commercialized exploitation of the mangrove forest are often outsiders, not the local people whose habitat is being destroyed (cf. Aksornkoae et al. 1984, 66).

Unless carefully controlled, the physical development processes (e.g. road-building) and large-scale logging or farming of forested land, both in the highlands and on the coast, may greatly reduce the ability of the forest to regenerate and may selectively change the character of the vegetation in favour of weedier (less useful) species. Over-cultivation in the highlands has sometimes resulted in replacement of forest by grasses (*Imperata* or *Saccharum*) or bracken that inhibit forest regeneration and greatly increase costs of further cultivation. The analogous development in over-used mangroves may be the spread of the fern *Acrostichum,* which inhibits the reseeding and regeneration of mangrove-tree species (Gan 1982; Srivastava and Khamis 1973; Srivastava and Sani 1979). Ecological problems (e.g. erosion) in the highlands are often associated with use of earth-moving machinery. In the mangrove forest, earth-moving projects which block water-flow patterns kill the mangroves as a result of change of water quality (see, e.g., Saenger, Hegerl, and Davie 1983, 35). Drying of the mangrove soils associated with some types of agricultural development may result in diminishing crop yields and virtually irreversible acid sulphate buildup (Hamilton and Snedaker 1984, 94–95; Saenger, Hegerl, and Davie 1983, 38).

There seems to be a strong association between commercialization of land use and costly, destructive ecological changes that generally do not occur under traditional subsistence uses. Great damage often occurs during the early stages of development, when the economy shifts from a sustainable subsistence system to an extractive commercial basis. In the highlands opium cultivators are generally much more destructive of forest resources and much less able to maintain yields

on a given site than are subsistence cultivators
(Kunstadter and Chapman 1978). In upland
forests, commercialized timber harvesting has
often been by high-grading (selective cutting of
valuable species) without replanting, leaving vast
tracts of low-quality forest. In the mangroves,
large-block cutting without systematic replanting
greatly increases the time needed for natural
reforestation. Also in the mangroves the commercialization of fishing often leads to over-fishing.
The problem of assessing the costs and controlling
the damage may be more difficult in the mangroves because the consequences are often geo-
graphically and socially remote from the places
where the decisions are made and the environmental interventions take place. It is the fishermen, not the mangrove developers, who suffer
directly from the loss of spawning grounds. This is
a special example of "the tragedy of the commons" — i.e. the difficulty of exerting control for
widespread public benefit over a resource that
may be exploited for localized gain (Bromley 1985;
Hardin 1973).

It is apparent that, once the pressures of modernization (e.g. in demands for cash and the material
goods money can buy) spread from outside to the
mangrove dwellers, the resulting changes take on
a momentum of their own. In seeking the better
things of modern life through commercialization,
the descendants of the traditional mangrove settlers themselves may actively participate in the
forces that destroy their environment. Community
ethics break down, local inhabitants are no longer
able to control the use and changes of their environment, the users (or despoilers) of the environment do not have to live with the immediate or
long-term consequences of the changes, and the
mangrove dwellers themselves can exert little or
no influence over these changes.

The overall rate at which mangrove forest is being
converted to other uses or made unsuitable for
continued forest use may be on the order of 2 to
8 per cent per year (Sabhasri 1982). This rate is
about the same as that for upland forests. If sustained, it leads to very rapid disappearance of the
forests. About 40 per cent of the mangrove area
of the Philippines was reported to have been converted to fish ponds between 1967 and 1978
(Hamilton and Snedaker 1984, 25–26; Saenger,
Hegerl, and Davie 1983, 36ff.). About 200,000
ha of mangroves were reclaimed for agriculture in
Indonesia in 1969–1974, with a much larger area
reclaimed in 1974–1979 (Bird and Ongkosongo
1980). The potential for continued loss of man-

groves is great because less than 1 per cent of
the total world mangrove area is under some
degree of preservation status (Hamliton and Snedaker 1984, 9). Because of the special problems
of agricultural development in coastal zones (e.g.
acid sulphate buildup), the "reclaimed" mangrove
areas may become unsuitable for agriculture after
a few years of cultivation.

Pressure on the remaining mangrove forest areas
is often very great because they may appear to be
relatively low-cost sites for construction of a variety of facilities, from salt farms, fish ponds, and
rice fields to housing estates and industrial sites.
If present socio-economic and demographic trends
outside the mangroves continue, these pressures
can be expected to increase.

Management of the Mangroves

Most of these problems are now well recognized,
and in recent years there has been a shift in
research emphasis from basic descriptive studies
to studies of management systems. Recent
studies have often proceeded from several basic
assumptions about the aims of management.
These include combining some of the virtues of
the traditional systems with the benefits of
modern use — long-term sustainable yield, multiple use, reduction of environmental damage, and
maintenance of the ecological benefits of the mangrove forest at the same time as economic goods
are being taken from them for human use — and,
in some areas, deliberate protection of mangroves
in a relatively unmodified state (Burbridge and
Koesoebiono 1980; Hamilton and Snedaker 1984;
Knox and Miyabara 1984, chs. 6 and 7; Nature
Conservation Council 1984; Philippine Council for
Agriculture and Resources Research 1978,
71–128; Saenger, Hegerl, and Davie 1983, ch. 6).
Management of mangroves may pose special
problems, as compared with other ecological systems, for reasons already given, especially because of the important linkages of mangroves to
other ecological zones.

Probably the most detailed integrated research on
ecology, traditional patterns of human use, and
modern management of mangrove areas for higher
yields has been in Indonesia (e.g. Collier 1979;
Hanson 1981; Hanson and Koesoebiono 1979;
Koescebiono et al. 1979). Indonesia is a particularly appropriate place for this research because of
the pressing need for agricultural land due to
rapid population increase. Indonesia's long-term

Characteristics of mangrove management systems

	Traditional systems	Transitional systems	Ideal developed systems
Population	Small; slow growth; little net migration	Rapid growth; net in-migration	Large; slow growth; little net migration
Technology	Simple; low use of machinery and chemicals	Increasing use of machinery and chemicals	High use of machinery and chemicals
Use of resources	Largely local	Increasingly national and international	Local, national, and international
Employment	Self-employed; local	Corporate; remote	Self-employed and corporate
Economic-system boundaries	Largely self-contained, involving trade and barter	National and international; commercial	National and international; commercial
Yield	Relatively low	Temporarily high, then declining	Moderate to high
Net productivity	Self-sustaining	Extractive	Self-sustaining, with inputs for restoration
Purposes	Multi-purpose	Often single-purpose	Multi-purpose
Knowledge used for management	Local, detailed, traditional	Technical; general	Scientific, local, detailed, and general
Management objectives	Subsistence in perpetuity	Profit	Profit and sustainability
Method of control of exploitation	Customary behaviour and values supported by local moral community	Poorly enforced laws and regulations; loss of moral community	National and international regulation, and international moral community (e.g. control of trade in endangered species)
Pollution	Local, biodegradable, chemically non-toxic, minor, micro-biological pollution may be effectively controlled by dilution	Local and regional; bio-degradable and non-biodegrable; non-toxic and toxic; major (oil; agricultural and industrial chemicals); poorly controlled, with danger of secondary spread by marketing	Full range of potential sources and types; actively controlled

transmigration programme has sought, with mixed success, to relieve population pressures in Java through government-sponsored migration to less-populated islands, especially Kalimantan and Sumatra. The destination has often been lands that are marginal for agriculture and poorly suited for traditional Javanese farming techniques. The work of scholars in Indonesia has described ecological problems, management issues, and some successful adaptations of traditional solutions, such as those of the Buginese, to coastal conditions. These studies can serve as a model for other areas in which there is a general lack of similar integrated research.

Discussion

Ecological characteristics of mangroves are in general fairly well known, but detailed information is needed on local and regional variations. This is important in discussing socio-economic aspects of human settlements because mangroves have hinterlands with a great diversity of natural and socio-economic environments which exert a strong influence on ecological processes and human activities within the mangroves. Mangrove areas are transitional zones and are thus affected in many ways by inland and seaward conditions. The traditional socio-economic systems that exploited mangrove resources were relatively small-scale and poor in technological equipment but often quite rich in intimate knowledge of the ecosystems. As economic development has advanced, technological influences over the mangrove environment have increased, but these have often been detrimental to the mangrove ecosystem. Except in Indonesia, few attempts have been made to describe traditional mangrove dwellers' knowledge and utilize it in designing management systems. This knowledge, together with the social organization by which it is implemented, is a valuable resource for reaching the management goals of sustained yield and multiple use.

In several respects the present socio-economic situation of traditional settlers in the mangroves resembles that of traditional marginal farmers of the upland forests. Both are now feeling the effects of large, scale socio-economic and demographic processes associated with modernization, urbanization, and socio-economic development, including commercialization of traditional activities, introduction of new demands for cash, loss of land, and damage to the ecosystem associated with development activities. Mangrove forests in Thailand were apparently being converted to other uses at about 6,350 ha per year between 1975 and 1979 (ch. 2, table 6, below). If conversion continues at the same rate, the mangroves will be reduced to about 75 per cent of their 1979 area by 1990, and only about 54 per cent will remain by the year 2000. Although there has been much discussion of the probable effects of the decline in mangrove areas on ocean fisheries and on forestry, to date little attention has been paid to the fate of the traditional mangrove dwellers when their forest has been heavily damaged or converted to other uses.

Likewise, although there has been considerable research on the effects of pollution on mangrove flora and fauna, there has been little attention paid to the biological or economic effects on human populations (e.g. with regard to direct toxic effects or to the saleability of mangrove products from contaminated sources). Another area that deserves further study is the effect of modification of mangroves on health conditions for diseases besides malaria (e.g., what are the consequences

of increased population density on traditional methods of waste disposal?).

One of the limitations on research on the socio-economic systems of the mangroves is the lack of basic information on who is living in the mangroves and what they are doing. Answers to these questions require field research, but it should also be possible to collect statistical information systematically if national governments geo-coded the data they routinely collect in censuses and other official surveys. Geo-coding (applying a designation to data to specify the geographical location from which it was collected, in addition to the administrative district) would allow detailed mapping of socio-economic and demographic data along with information which is already mapped, such as the distribution of mangrove forests or other ecological zones (see, e.g., Central Bureau of Statistics 1979; Prasartkul et al. 1980; Salih 1979).

The mangrove ecosystem and the socio-economic systems of mangrove residents are both greatly affected by processes and events beyond the geographical borders of the mangrove forests. This implies that it is appropriate for planners to take a broader than usual view when planning for the development of mangrove areas or when regulating uses of the mangroves. It also means that the ability to influence the future of the mangroves in the direction of sustained multiple use may be limited unless controls can be put on pollution and on economic-development activities that affect the mangroves and their human populations. This may require the development of new community ethics with sufficient social and geographical scope to encompass both the social systems in which decisions affecting environmental interventions in the mangroves are made and the mangrove zones themselves.

References

Aksornkoae, Sanit, Somsak Priebprom, Anant Saraya, Jitt Kongsangchai, Puckprink Sangdee, et al. 1984. *Research on the socio-economics of dwellers in mangrove forests, Thailand.* Faculty of Forestry, Kasetsart University, Bangkok, Thailand.

Arwati. 1977. *Kasub did malaria* [Malaria control]. Cross-sectoral background paper based on malaria control technical study. Ministry of Health, Jakarta.

Bird, Eric C. F., and Otto S. R. Ongkosongo. 1980. *Environmental changes on the coasts of Indonesia.* United Nations University, Tokyo. (NRTS-12/UNUN-197)

Boschi, L. 1975. *Field visit to Burma (Arakan and Tenasserim coastal area) for mosquito control: Report on a preliminary* survey (21 October—11 November 1975). WHO Project SE ICP MPD 001 (SEARO 007). World Health Organization, New Delhi.

——. 1976. Report on training course on comprehensive malaria control operations (Ngapali – Sandoway – Burma), 17 January to 5 February 1976. WHO Project SE ICP MPD 001 (SEARO 007). World Health Organization, New Delhi.

Bromley, Daniel W. 1985. "Common property issues in international development." BOSTID Developments (Board on Science and Technology for International Development, National Research Council, Washington, D.C.), 5 (1): 12–15.

Burbridge, Peter R., and Koesoebiono. 1980. Management of mangrove exploitation in Indonesia. Pusat Studi Pengelolaan Sumberdaya dan Lingkungan (Centre for Natural Resource Management and Environmental Studies), Bogor University, Bogor, Indonesia. (PSPL/Research Report/007)

Central Bureau of Statistics, Republic of Indonesia. 1979. User's guide — Kabupaten/Kotamadya Data Bank (computerized integrated data file). 3rd ed. Central Bureau of Statistics, Jakarta.

Chia Thia-eng and Joseph K. Charles, eds. 1984. Coastal resources of east coast Peninsular Malaysia: An assessment in relation to potential oil spills. Penerbit Universiti Sains Malaysia, Penang, Malaysia.

Christensen, Bo. 1978. "Primary production of mangrove forests." Proceedings, International Workshop on Mangrove and Estuarine Area Development for the Indo-Pacific Region, 14–19 November 1977, Manila, pp. 131–135. Philippine Council for Agriculture and Resources Research, Los Baños, Philippines.

Collier, W. L. 1979. Social and economic aspects of tidal swamp land development in Indonesia. Development Centre Occasional Paper 15. Australian National University, Canberra.

Committee on the Effects of Herbicides in Vietnam. 1974. The effects of herbicides in South Vietnam. Part A, Summary and conclusions. National Academy of Sciences, Washington, D.C.

Desowitz, Robert S., S. J. Berman, D. J. Gubler, C. Harinasuta, P. Guptavanij, and V. Vasuvat. 1974. "Epidemiological-ecological effects: Studies on intact and deforested mangrove ecosystems." The effects of herbicides in South Vietnam. Part B, Working papers. National Academy of Sciences, Washington, D.C.

Gan, B. K. 1982. The role of Acrostichum aureum in the natural regeneration of mangrove trees, especially Rhizophora spp. in Malaysia. B.Sc., Forestry, thesis. Universiti Pertanian Malaysia, Serdang, Malaysia.

Gomez, E. D. 1980. The present state of mangrove ecosystems in Southeast Asia and the impact of pollution — Philippines. South China Seas Fisheries Development and Coordinating Programme, FAO and United Nations Development Programme, Manila. (SCS/80/WP/94c [Revised])

Hamilton, Lawrence S., and Samuel C. Snedaker, eds. 1984. Handbook for mangrove area management. United Nations Environment Programme, and Environment and Policy Institute, East-West Center, Honolulu, Hawaii, USA.

Hanson, A. J. 1981. "Transmigration and marginal land development." In G. E. Hansen, ed., Agricultural and rural development in Indonesia, Westview Press, Boulder, Colo., USA.

Hanson A. J., and Koesoebiono. 1979. "Settling coastal swamplands in Sumatra: A case study of integrated resource management." In C. MacAndrews and Chia Lin Sien, eds., Developing economies and the environment: The Southeast Asian experience, pp. 121–178. McGraw-Hill International Book Co., Singapore.

Hardin, Garrett. 1973. "The tragedy of the commons." In H. E. Daly, ed., Toward a steady-state economy. Freeman, San Francisco, Calif., USA.

Knox, George A., and Tetsuo Miyabara. 1984. Coastal zone resource development in Southeast Asia, with special reference to Indonesia. Unesco Regional Office for Science and Technology for Southeast Asia, and Resource Systems Institute, East-West Center, Honolulu, Hawaii, USA.

Koesoebiono, C. Maluk, G. W. Wiggin, and A. T. Hanson. 1979. "Resource use interaction in the mangrove forests of the Musi-Banyuasin coastal zone of South Sumatra." In A. R. Librero and W. L. Collier, eds, Economics of aquaculture, sea-fishing and coastal resource use in Asia, pp. 317–322. Agricultural Development Council and Philippine Council for Agricultural Resources, Los Baños, Philippines.

Kunstadter, Peter. In press. "Management of man, malaria and mosquitos in the mangroves." In P. Kunstadter and S. Snedaker, eds., Proceedings of Unesco Regional Conference on Mangroves, Dacca, December 1978.

Kunstadter, Peter, E. C. Chapman, and Sanga Sabhasri, eds. 1978. Farmers in the forest. University Press of Hawaii, Honolulu, Hawaii, USA.

Kunstadter, Peter, Sanga Sabhasri, Sanit Aksornkoae, Kasem Chunkeo, and Sathit Wacharakitti. In press. "Impacts of economic development and population change on Thailand's forests." In José I. Furtado anjd Kenneth Ruddle, eds., Tropical resource ecology and development.

Kunstadter, Peter, and K. K. Tiwari. 1977. Consultants' report on human uses and management of the mangrove environment in South and Southeast Asia. Unesco, New Delhi.

Lai Hoi-chaw and Feng Meow-chan, eds. 1984. Fate and effects of oil in the mangrove environment. Universiti Sains Malaysia, Penang, Malaysia.

MacNae, William. 1974. Mangrove forests and fisheries. Indian Ocean Programme, Indian Ocean Fishery Commission, FAO, and United Nations Development Programme, Rome. (IOFC/DEV/74/34)

Martosubroto, P., and N. Naamin. 1977. "Relationship between tidal forests (mangroves) and commercial shrimp production in Indonesia." Marine Research in Indonesia, 18: 81–86.

Nature Conservation Council. 1984. Strategies for management of mangrove forests in New Zealand: A discussion document prepared by a task force of the Nature Conservation Council. Nature Conservation Council, Wellington North, New Zealand.

Pejaranonda, Chintana. 1985. Declines in fertility by district in Thailand: An analysis of the 1980 census. Asian Population Studies Series, no. 62A. Economic and Social Commission for Asia and the Pacific, Bangkok.

Petr, T., ed. 1977. Purari River Environmental Studies (1). Workshop organized by Office of Environment and Conservation, 6 May 1977. Waigani, Papua New Guinea.

Philippine Council for Agriculture and Resources Research. 1980. Proceedings, International Workshop on Mangrove and Estuarine Area Development for the Indo-Pacific Region, Manila, Philippines, 14–19 November 1977. Philippine Council for Agriculture and Resources Research, Los Baños, Philippines.

Piyakarnchana, Twesukdi. 1980. The present state of mangrove ecosystems in Southeast Asia and the impact of pollution — Thailand. South China Seas Fisheries Development and Coordinating Programme, FAO and United Nations Development Programme. Manila. (SCS/80/WP/94e [Revised])

Prasartkul, Pramote, et al. 1980. "Towards the development of a territorial indicator system for Thailand." Paper prepared for National Statistical Office of Thailand Workshop on

Territorial Indicator Systems for Development Planning in Southeast Asia, Pattaya, Thailand, 12–14 May 1980, supported by Resource Systems Institute, East-West Center, and Ford Foundation. Institute for Population and Social Research, Mahidol University, Bangkok.

Sabhasri, Sanga. 1979. "Mangrove forest: A forest that grows in the sea." Paper prepared for George Long Lectureship, University of Washington. Seattle, Wash., USA.

——. 1982. "Mangrove ecology: A model of interdisciplinary research." In *Proceedings, National Research Council of Thailand – Japan Society for the Promotion of Science Ratanakosin Bicentennial Joint Seminar on Science and Mangrove Resources,* pp. 30–32. National Research Council of Thailand, Bangkok.

Saenger, P., E. J. Hegerl, and J. D. S. Davie, eds. 1983. *Global status of mangrove ecosystems.* Commission on Ecology Papers, No. 3. International Union for Conservation of Nature and Natural Resources, Gland, Switzerland.

Salih, Kamal. 1979. "Project NIDAS: Development of an integrated data system in Malaysia." Paper prepared for International Seminar on Information Systems in Public Administration and Their Role in Social and Economic Development, Chamrousse, France, 17–21 June 1979, organized by the Data for Development Association. Centre for Policy Research, Universiti Sains Malaysia, Penang, Malaysia.

Smithurst, N. A. 1970. "A malaria survey on Bougainville Island (Kieta Subdistrict) January–February 1967." *Southeast Asia Journal of Tropical Medicine and Public Health,* 1 (1): 117–122.

Soegiarto, Aprilani. 1980. *The present state of mangrove ecosystems in Southeast Asia and the impact of pollution — Indonesia.* South China Seas Fisheries Development and Coordinating Programme, FAO and United Nations Development Programme, Manila. (SCS/80/WP/94a [Revised])

Spenser, M. 1964. *Malaria in the d'Entrecasteaux Papua, with particular reference to Anopheles farauti Laveran.* World Health Organization, Geneva. (WHO/MAL/454)

——. 1976. *Age-grouping anopheline female populations with particular reference to Anopheles farauti no. 1 Laveran 1906 revised by Bryan 1973 (Diptera: Culicidae) in Papua New Guinea.* World Health Organization, Geneva. (WHO/MAL/76.865)

Srivastava, P. B. L., and D. Khamis. 1978. "Progress of natural regeneration after final felling under the current silvicultural practices in Matang Mangrove Reserve." *Pertanika,* 1 (2): 126–135.

Srivastava, P. B. L., and A. Sani. 1979. "Effect of final felling on natural regeneration in Rhizophora dominated forests of Matang Mangrove Reserve." *Pertanika,* 2 (1): 34–42.

Zoology Department, University of Singapore. 1980. *The present state of mangrove ecosystems in Southeast Asia and the impact of pollution — Singapore.* South China Seas Fisheries Development and Coordinating Programme, FAO and United Nations Development Programme, Manila. (SCS/80/WP/94d [Revised])

2. MANGROVE RESOURCES AND THE SOCIO-ECONOMICS OF DWELLERS IN MANGROVE FORESTS IN THAILAND

Sanit Aksornkoae, Somsak Priebprom, Anant Saraya, and Jitt Kongsangchai

Mangrove Resources in Thailand

Definition

Mangrove forests are one of the primary features of tropical and subtropical coastal ecosystems. They reach their maximum development in South-East Asia. Definitions of mangrove forest generally embody two different concepts: The first refers to an ecological group of evergreen plant species belonging to several botanical families but possessing marked similarities in their physiological characteristics and structural adaptation and having similar habitat preferences. The second concept implies a complex of plant and animal communities. In Thailand the mangrove plant community is composed mainly of *Rhizophora* species associated with other trees and shrubs growing in the zone of tidal influence, both on the more sheltered parts of the coast and inland, along river banks and tidally flooded parts of estuaries.

Distribution of Mangrove Forests

Mangrove forests in Thailand occur on seashores, around lagoons, and along rivers at levels between low and high tides, in the southern and south-eastern parts of the country and the upper part of the Gulf of Thailand (fig. 1). The extent of existing mangrove forest in 1979, as estimated from satellite imagery, was approximately 287,308 ha, or 1,795,675 *rai* (Klankamsorn and Charuppat 1982). A breakdown of mangrove areas by province is shown in table 1. Out of the total, 176,653 ha (61.5 per cent) is managed by the Royal Forest Department. The remaining areas are private property, and some are classified as unproductive forests.

Species and Zonation

Table 2 lists mangrove species found in Thailand, including 27 genera of trees and other plants.

This study was supported by the United Nations University.

Rhizophora species are the most abundant and have the widest geographical distribution.

In most mangroves, different species dominate certain zones. The characteristic zonation pattern results from differences in the rooting and growth of seedlings and the competitive advantage each species has along the gradient from mean sea level to above the high-water line. Aksornkoae (1976) studied the dominant species associations

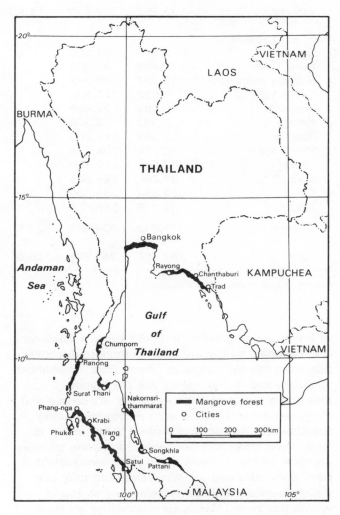

FIG. 1. Distribution of mangroves in Thailand

TABLE 1. Area of mangrove forests in Thailand by province, estimated from 1979 Landsat satellite imagery

Region and province	Area		% of total
	ha	rai	
South-eastern Thailand			
Cholburi	3,312	20,700	
Rayong	4,608	28,800	
Chanthaburi	24,064	150,400	
Trad	9,840	61,500	
subtotal	41,824	261,400	14.6
Upper Gulf of Thailand (southern Chao Phya plain)			
Chachoengsao	2,320	14,500	
Samutprakarn	1,040	6,500	
Samutsakorn	14,416	90,100	
Samutsongkram	7,648	47,800	
Phetchaburi	7,792	48,700	
Prachuabkhirikan	336	2,100	
subtotal	33,552	209,700	11.6
Southern Thailand			
Chumporn	6,928	43,300	
Surat Thani	5,808	36,300	
Nakornsrithammarat	12,832	80,200	
Pathalung	1,632	10,200	
Songkhla	5,184	32,400	
Pattani	1,392	8,700	
Ranong	22,592	141,200	
Phang-nga	48,716	304,475	
Phuket	2,848	17,800	
Krabi	31,760	198,500	
Trang	32,864	205,400	
Satul	39,076	246,100	
subtotal	211,932	1,324,575	73.8
Total	287,308	1,795,675	100.0

Source: Klankamsorn and Charuppat.

TABLE 2. Mangrove species of Thailand

Family	Species
Rhizophoraceae	*Rhizophora apiculata, R. mucronata, Bruguiera cylindrica, B. parviflora, Ceriops roxburghiana, Kandelia rheedii*
Sonneratiaceae	*Sonneratia caseolaris, S. alba*
Verbenaceae	*Avicennia alba, A. officinalis, Clerodendrum inerme*
Meliaceae	*Xylocarpus obovatus, X. moluccensis*
Myrsinaceae	*Aegiceras corniculatum*
Caesalpiniaceae	*Caesalpinia didyna, Intsia retus*
Palmae	*Nypa fruticans, Phoenix paludosa*
Sterculiaceae	*Heritiera littoralis*
Combretaceae	*Lumnitzera racemosa*
Myrtaceae	*Melaleuca leucadendron*
Apocynaceae	*Cerbera odollam*
Acanthaceae	*Acanthus ebracteatus, A. ilicifolius*
Euphorbiaceae	*Excoecaria agallocha*
Pteridaceae	*Acrostichum aureum, A. speciosum*
Rubiaceae	*Litosanthes biflora, Scyphiphora hydrophyllacea*
Malvaceae	*Hibiscus tiliaceus, Thespesia populnea*
Lauraceae	*Cassytha filiformia*
Ebenaceae	*Diospyros ferrea*
Flagellariaceae	*Flagellaria indica*

of mangrove vegetation in south-eastern Thailand and summarized the variation in mangrove vegetation from the river edge to inland sites. *Rhizophora apiculata* and *R. mucronata* are the dominant species along river and channel banks. *Avicennia* and *Bruguiera* are associated with *Rhizophora* along the channels, but form a distinct zone further inland. *Xylocarpus* and *Excoecaria* dominate on sites adjacent to the *Avicennia* and *Bruguiera* zone that have drier soils and are less subject to tidal inundation; *Ceriops* and *Lumnitzera* are also found within this zone. *Melaleuca* reaches its greatest dominance further inland on even drier and more

elevated sites that are still less subject to tidal flooding. The fern *Acrostichum aureum* occurs throughout the mangrove area but is densest in sites that have been disturbed.

Human Settlements and Populations in Mangrove Areas

Thailand has a total of about 2,600 km of coastline, of which 927 km are lined with mangroves (Klankamsorn and Charuppat 1982). People who live in or earn their living from the mangroves use the zone for timber, poles, fuelwood, and thatch, for capture and culture fisheries, for salt production, and for other purposes.

It is difficult to estimate how many people are dependent on this resource. Unpublished data collected by the Royal Forest Department show the average number of persons working in the concession forests is 3.7 per km^2 and the number of people living in and adjacent to the mangroves is about 34 per km^2. It is estimated that about 98,000 people live within and near the 287,308 ha of mangroves. Of these, 8,375 work in the 176,653 ha of concession forests.

Most mangrove dwellers live in clustered houses in small village communities at the edge of the forest or along channels within an estuary. Most of the villagers are of Thai origin; other ethnic groups include Muslims speaking a Malay dialect, and Chinese.

The villages can be classified into three main types: island villages, shrimp-pond villages, and coastal-land villages. Most of these villages comprise 20–50 households, with populations of 100–300, but larger villages may incorporate more than 100 households and a population of 600 (Petchmedyai 1980).

Traditional Uses of Mangrove Resources

Forest Products

The traditional uses of mangrove wood in Thailand are for charcoal and firewood, poles and construction materials, fishing gear, tanning, and medicines. In addition, there has been some distillation of chemical products from mangrove wood. The palm *Nypa fruticans* provides materials for several purposes.

Charcoal. About 90 per cent of the wood harvested from mangrove forests is used for charcoal production. Various species of the Rhizophoraceae can be used to make charcoal, but *Rhizophora apiculata* and *R. mucronata,* which produce heavy, dense, hard charcoal with a high calorific value that is almost smokeless when burned, are preferred. Other species, such as *Bruguiera* spp.

and *Ceriops* sp., are also used but in minimal quantity. The minimum size of stems for charcoal burning is about 5 cm. The average harvest is about 783,780 m³ of wood per year, which produces about 387,800 m³ of charcoal — or about 263,704 metric tons on the basis of 680 kg per cubic metre (Phuritat 1975). This charcoal is produced in 1,273 kilns located in or near the mangrove areas. Typical charcoal kilns are dome-shaped, made of clay and bricks, and have a capacity of between 50 and 200 m³. The average amount of wood used, number of kilns, and charcoal production in different parts of the country under five regional forest management offices are shown in table 3. Charcoal produced from mangrove wood is used in Bangkok and coastal towns, in addition to which a surplus is exported to neighbouring countries.

Firewood. Wood from mangrove forests is also widely used by mangrove dwellers and people who live along the coastline as firewood. No data are available on the amount of firewood consumed per year per species. The mangrove species commonly used for firewood are *Avicennia, Xylocarpus, Excoecaria, Bruguiera,* and *Lumnitzera* spp. *Rhizophora* spp. are also used for firewood but only in small quantity because these species are preferred for charcoal.

Poles. Mangrove poles are used for many purposes where water-resistance is needed — especially for foundation pilings, ore-rinsing troughs, and fishing stakes. The quantity of wood used for these purposes is minimal, however, and no data are available on the total amount. Mangrove species commonly used for poles are *Rhizophora apiculata, R. mucronata, Ceriops* sp., *Bruguiera*

TABLE 3.   Production of charcoal from mangrove wood in Thailand, 1978–1982

Regional forest	Mangrove area under management (ha)	Harvesting area[a] (ha/yr)	Wood production (m³/yr)	Number of kilns	Capacity per kiln (m³)	Burning cycles per year	Charcoal production (m³/yr)
Suratthani	30,786	1,026	198,806	249	50–200	5–10	95,258
Nakorn-srithammarat	72,212	2,407	382,905	540	100–180	6–7	191,507
Songkla	51,041	1,701	181,886	451	120–170	4–5	90,943
Pattani	752	25	2,785	4	110	7	1,393
Sriracha	21,862	729	17,398	29	70–195	5–6	8,699
Total	176,653	5,888	783,780	1,273			387,800

Source: Royal Forest Department.

a. With 30-year rotation.

spp., and *Excoecaria agallocha. Excoecaria agallocha* seems to be more widely used than the others.

Construction materials. Mangrove lumber is used for house construction mainly by people who live in or close to the mangrove forest. Wood of various mangrove species can be used for different parts of the house. *Rhizophora, Avicennia, Bruguiera,* and *Xylocarpus* are commonly used for columns, bracing members, beams, and roof frames. Floors and platforms are made only of *Rhizophora* and *Bruguiera* species. Unfortunately the quantity of mangrove wood used for construction has not been recorded.

Fishing gear. Various types of fishing gear are used by mangrove dwellers, and some of the equipment is made of mangrove wood. Poles from *Rhizophora* spp. are usually used for crab traps. The crab most commonly caught in the mangrove area is *Scylla serrata.* Drift gill nets and winged set bags also use mangrove poles; *Rhizophora apiculata* and *Bruguiera* spp. are the species commonly used for this purpose. The volume of wood used for fishing gear is small.

Tanning. Mangrove bark is now only rarely used for tanning in Thailand. Formerly tannin from *Rhizophora* and *Ceriops* was used by fishermen for dying their fishing nets, but now they mainly use nylon nets, and little tannin is used for dyeing nets.

Medicines. Mangrove dwellers in Ranong Province, on the west coast, have reported that they commonly use various mangrove species for medicines. *Acanthus ebracteatus* and *A. ilicifolius* are used to treat kidney stones, *Bruguiera parviflora* is used to relieve constipation, *Avicennia alba* and *A. officinalis* are used to treat thrush in children, and *Triumfetta rhomboidea* is used for menstrual fevers (see ch. 3, below, for details).

Distillation. The only plant for wood distillation in this part of the world is at Kapur, near Ranong. Raw distillate from mangrove wood (*Rhizophora apiculata*), consisting essentially of pyroligneous acid, is collected from the vents of a charcoal kiln by condensation. This condensate can be fractionated by a complicated process, producing 5.5 per cent acetic acid, 3.4 per cent methanol, and 6.5 per cent wood tar. At present, however, problems with the extraction technique make the process uneconomical.

Nypa fruticans. The nipa palm, *Nypa fruticans,* which is widely distributed in the mangrove area, is used for a great variety of purposes, the chief one being the use of its leaves for thatching. Nipa leaves are also popularly used locally for cigarette wrappers. Elsewhere, alcohol is extracted from nipa sap, but this is not done in Thailand.

Fishery Products

Fish production includes capture and culture fisheries.

Capture fishery. More than 1.7 million metric tons of marine fish and crustaceans were caught in Thailand from 1979 through 1980(table 4). Of this amount, almost 120,000 tons were shrimp, which fetch the highest price per kilogram and are often most abundant in or adjacent to mangrove areas. Of the total shrimp production, about 6 per cent was from coastal aquaculture (much of it in mangrove zones), 22 per cent from small-scale fishing (mostly mangrove capture fishery), and 72 per cent from large-scale fishing (mainly off-shore operations). Many of the shrimp from large-scale fishery are mangrove-dependent species (e.g. *Penaeus merguiensis, Metapenaeus* spp.). The degradation or reduction of mangrove areas could adversely affect the coastal fisheries, particularly the capture of shrimp (Martosubroto and Naamin 1977; Gedney, Kapetsky, and Kuhnhold 1982).

Culture fishery. Of the total marine shrimp production in Thailand in 1979, about 6 per cent, 7,000 tons, was from coastal aquaculture (see table 5). Most of the aquaculture farms in Thailand, particularly shrimp farms, are in the mangrove forest zone in Samutsakorn, Samutsongkram, Samutprakarn, Surat Thani, and Nakornsrithammarat provinces. About 62,400 ha of the total mangrove forest area in Thailand is suitable for shrimp aquaculture (National Research Council of Thailand 1977). The 7,000 tons of shrimp produced in

TABLE 4. Marine fishery production in Thailand, 1974–1980 (metric tons)

	Total production	Marine shrimp
1974–1976	1,432,663.3	85,268.0
1977–1978	2,012,659.0	123,175.0
1979–1980	1,730,555.5	117,398.5
Total	5,175,877.8	325,841.5

Source: Chaitiemvong 1983.

TABLE 5.   Shrimp culture production in Thailand, 1974–1979

	Number of shrimp farms	Area (ha)	Production (metric tons)			
			Penaeus spp.	*Metapenaeus* spp.	Other species	Total
1974	1,518	12,092	862	697	234	1,793
1975	1,568	12,867	1,179	1,121	237	2,537
1976	1,536	12,296	1,275	1,009	239	2,523
1977	1,438	12,418	721	599	270	1,590
1978	3,045	21,730	4,117	1,650	628	6,395
1979	3,378	24,675	5,048	1,359	665	7,072

Source: Chaitiemvong 1983.

TABLE 6.   Mangrove destruction in Thailand, 1961–1979

	Mangrove area (ha)	Extent of destruction			
		Total		Average per year	
		ha	%	ha	%
1961	367,900[a]				
		55,200	15.0	3,943	1.07
1975	312,700[b]				
		25,392	8.1	6,348	2.03
1979	287,308[c]				
1961–1979		80,592	21.9	4,477	1.21

a. Sukwong et al. 1976.
b. Vibulsresth, Ketruangrote, and Sriplung 1976.
c. Klankamsorn and Charuppat 1982.

1979 came from 3,378 farms, occupying a total area of about 24,675 ha. Production was thus only about 284 kg per hectare. This low figure suggests that most of the farms are not very productive. In general they have been found to be productive only for the first three or four years of operation and are often abandoned after that. This practice is inefficient and very wasteful, resulting in the degradation or destruction of large areas of mangroves, and may in turn adversely affect the yield of coastal shrimp capture.

The Destruction of Mangroves

Factors Leading to Destruction

Population increase, aquaculture, mining, salt ponds, and the over-exploitation of mangrove forests without replanting are important factors leading to the progressive destruction of mangrove forests.

Rapid increase of population. During the years 1950–1981, the population of Thailand grew at an average annual rate of about 2.7 per cent. The rate of increase is now below 2 per cent as a result of family planning programmes. Population growth led to urban expansion and to increased demand for agricultural land. Since upland and lowland areas were already occupied, large numbers of people have moved to coastal areas, including mangrove forest areas, but no data are available on the size of the areas affected.

Aquaculture. Aquaculture, especially shrimp and fish farming, is widely practised in the mangrove area. Shrimp and fish ponds have been built by clear-cutting the mangrove forest, levelling, and building embankments. The area of shrimp farms increased from 12,867 ha in 1,568 farms in 1975 to 24,675 ha in 3,378 farms in 1979 (table 5), not including areas converted to aquaculture and then abandoned without replanting.

Mining. Hydraulic tin mining in the mangrove area has been carried out mostly in Ranong, Phang-nga, and Phuket provinces. Mining results in the deposition of large quantities of silt in stream channels and in the inter-tidal and near-shore zones. The extent of tin mines in the mangrove forest was approximately 926 ha in 1979 (Klankamsorn and Charuppat 1982), but the area affected by siltation from tin mines was much larger than this.

Salt ponds. Large areas of mangrove forest, totalling about 10,356 ha, have been converted to salt ponds. Most of these are located in the upper parts of the Gulf of Thailand.

Over-exploitation for forest products without replanting. In various parts of the mangrove forests, trees have been cut at a rate beyond their ability to regenerate. After cutting, some mangrove areas have been left without replanting.

Rate of Destruction

During the period 1961–1979 the total area of mangrove forest converted to various purposes was about 80,592 ha (Klankamsorn and Charuppat 1982). Table 6 shows that the rate of destruction of mangrove forests was high from 1975 to 1979, about 6,348 ha per year, as compared with 3,943 ha per year from 1961 to 1975.

Case Study: The Socio-economics of Mangrove Forest Dwellers in Ranong Province

The rapid decline and deterioration of mangrove ecosystems in Thailand under stress from population growth and economic development, to which scientists and conservationists have called attention, implies the need to recognize the importance of conservation, management, and restoration of mangrove forests so that productivity may be optimized and the best economic yield obtained on a long-term basis without destroying the mangrove ecosystem. Several organizations, especially the Royal Forest Department, have encouraged scientists to develop management plans for the maximum and efficient use of mangrove resources. Future management plans for mangrove forests will also emphasize the improvement of the quality of life of mangrove dwellers.

One major constraint on development of a mangrove management plan for improving the quality of life of mangrove dwellers is lack of basic

knowledge about community structure, resource utilization, and economic conditions of people living in mangrove communities. It is therefore necessary to make intensive studies of mangrove settlements, which depend mainly on the productivity of mangrove organisms. Decisions on the future use of mangrove ecosystems based on inadequate knowledge of these people may result in unanticipated hardship for mangrove dwellers, and irrevocable loss of valuable mangrove resources.

To obtain basic knowledge of mangrove communities, the United Nations University supported a research project on the socio-economics of dwellers in mangrove forests. Field study was carried out from September 1983 to June 1984 in the villages of Ko Lao and Had Sai Khao, both in Ranong Province in southern Thailand.

Research Objectives

The main objectives of the research project were:
— to study the sociological characteristics of mangrove dwellers,
— to study the household economics of mangrove dwellers,
— to evaluate socio-economic changes and the quality of life of mangrove dwellers,
— to study and evaluate differences in life-styles in communities of mangrove dwellers with different cultural-religious backgrounds, and
— to provide baseline data for developing a mangrove management plan in order to improve the traditional way of life of mangrove dwellers.

Methodology

Selection of the study area. Ko Lao and Had Sai Khao villages were chosen as study sites for several reasons. They provide good examples of different groups of mangrove dwellers. In Ko Lao the villagers include both Muslims and Buddhists, while in Had Sai Khao they are exclusively Buddhists. Each village has 26 households. The villages are about 4 km apart, on the shores of the Ranong estuary.

Collection of data in the field. Data were gathered by means of interviews. A questionnaire and a survey of socio-economic characteristics included household size and composition by age, sex, education, occupation, and religion; mobility and land tenure; consumption of forest products; fishery production; and health and sanitary conditions. Representatives of each household, generally the heads of the families, were interviewed by graduate students who lived in the villages for various periods. Great care was taken to establish rapport and to convince the interviewed persons that the results of this study would be used in revising the mangrove management plans and other mangrove activities and that this would improve their traditional way of life. Our impression is that the interviewed persons provided as accurate information as they possibly could.

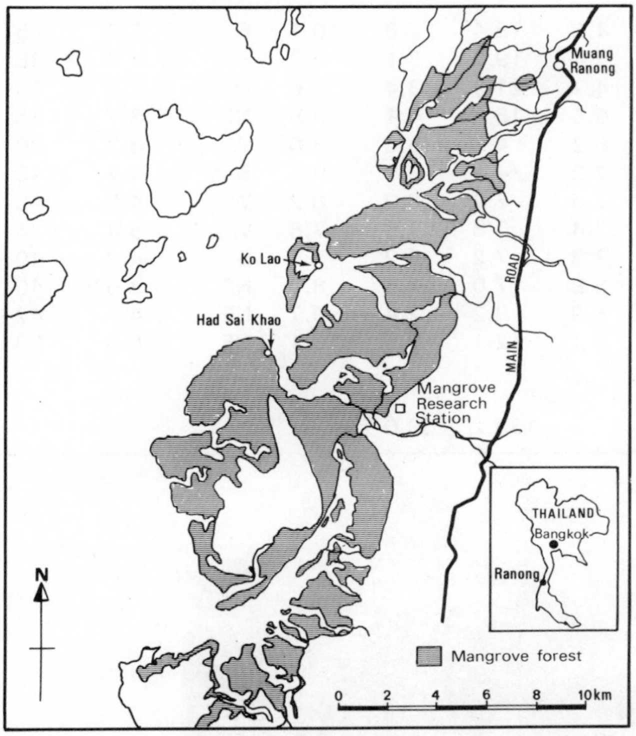

Data analysis. Numerical and percentage frequency distributions were computed from field survey data for each village in order to describe socio-economic characteristics such as age, sex, education, and occupation. Data on economic conditions were analysed in relation to sources of income and expenditures. Other information such as health and sanitary condtions, daily life, and perceptions regarding use and conservation of mangrove resources were also analysed. These tabulations allowed comparison of life-styles of population groups with different religions.

General Description of the Study Area

Location. Ko Lao and Had Sai Khao villages are located at latitude 9°50′ N and longitude 98°35′ E, about 10 km south-west of Ranong city and about 650 km south of Bangkok (fig. 2). Ko Lao is in Paknam commune (*tambol*) and Had Sai Khao is in Ngao commune, both in Muang district (*amphoe*) in Ranong Province (*changwat*). The two villages are located along the coastline of the Ranong estuary, fringed by mangrove forests (fig. 3).

FIG. 2.   Location of Ko Lao and Had Sai Khao villages

FIG. 3.   Ko Lao village, Ranong Province, Thailand

TABLE 7.   Climatological data from Ranong Province, 1951–1975

	Temperature (°C)			Monthly rainfall (mm)	Rainy days	Relative humidity (%)	Cloudiness (okta)	Days with haze	Days with fog	Visibility (km)	Prevailing wind	Wind velocity (knots)	
	Max.	Avg.	Min.									Mean	Max.
January	31.5	25.8	20.3	29.9	5.2	76	4.5	18.4	4.6	10.1	E	5.5	25
February	33.0	26.9	20.6	16.8	3.6	75	4.4	19.2	3.1	9.9	E	4.9	26
March	34.0	28.1	21.7	47.6	6.0	74	4.7	21.3	3.4	9.6	ENE	4.4	30
April	34.0	28.3	23.0	153.9	12.7	78	5.5	15.5	4.1	9.8	NE	3.7	35
May	31.8	27.2	23.4	473.5	22.6	85	6.9	4.3	5.6	9.0	W	3.3	30
June	30.2	26.4	23.3	765.6	26.4	88	7.3	4.9	3.9	8.1	S.W	4.7	40
July	30.1	26.2	23.1	686.1	26.4	88	7.3	7.4	3.2	8.2	W	4.5	45
August	29.8	25.9	23.1	786.9	27.4	89	7.4	7.8	1.5	7.8	W	5.4	35
September	29.8	25.7	22.9	708.3	26.0	89	7.3	7.2	2.2	7.7	S	4.2	40
October	30.3	25.9	22.5	430.9	24.1	89	6.8	7.0	2.5	8.9	NE	3.6	40
November	30.5	25.8	21.9	168.5	15.9	85	6.9	8.2	2.4	9.7	NE	4.1	47
December	30.5	25.5	20.9	52.5	7.7	79	5.1	12.1	2.5	10.1	NE	5.6	28
Annual total				4,320.3	204.0			133.2	39.0				

Source: Sarigabutr et al. 1982.

FIG. 4.   *Rhizophora* trees along the edge of the estuary near Ko Lao and Had Sai Khao villages

Climate. The climate is humid tropical. There is little variation in mean monthly temperature, with the values between 25.5° and 28.3°C. Annual rainfall is about 4,320 mm, the highest in Thailand. Most of the precipitation falls during the rainy season, from April to November. There are about 204 rainy days per year. A summary of climatological data for Ranong Province between 1951 and 1975 is shown in table 7.

The mangrove forest. A Mangrove Research Station was established in 1983 about 4 km from Ko

FIG. 5.   *Avicennia* and *Sonneratia* trees along the edge of the estuary

Lao and 3 km from Had Sai Khao, and since then the forests have been reserved for research. One concession for forestry, based at Ngao, about 10 km from the research area, had been granted before the research station was established. That area is now reserved for ecological studies of the mangrove ecosystem. Because of selective cutting in this mangrove forest for many years in the past, the forest is composed mainly of small, closely spaced trees. The average stem volume of this forest is only about 10 m³ per ha (Kooha 1983). Only parts of the forest, particularly those near Had Sai Khao village, are composed of large trees, especially *Rhizophora apiculata* and *Bruguiera gymnorrhiza,* with a diameter at breast height exceeding 60 cm and an average stem volume of about 120 m³ per ha (Jintana et al. 1983) (fig. 4).

The mangrove forests around Ko Lao and Had Sai Khao generally show a diversity of species with clear zonation (Aksornkoae et al. 1982; Miyawaki et al. 1983). *Rhizophora* spp., especially *R. apiculata* and *R. mucronata,* grow mainly along side creeks. Locally, along the seaward edge where the soil is sandier, the *Rhizophora* community is found behind the *Avicennia* and *Sonneratia* zones (fig. 5). The common species of *Avicennia* are *A. alba* and *A. officinalis. Sonneratia alba* is present in this area rather than *S. caseolaris.*

Bruguiera spp., especially *B. cylindrica* and *B. parviflora,* are commonly found behind the *Rhizophora* community. *Ceriops,* particularly the *C. tagal* community, grows on more elevated inland sites, and *Xylocarpus,* especially *X. granatum,* is found scattered within the *Ceriops* community.

Research Results and Discussion

Forestry

The three main uses for mangrove wood by residents of Ko Lao and Had Sai Khao are as fuel (firewood and charcoal), for fishing gear, and for house construction.

Fuel

Most Ko Lao villagers (92 per cent) used firewood for cooking, and about 58 per cent supplemented it with charcoal (fig. 7). The average household consumption of fuelwood (firewood and charcoal) in Ko Lao was estimated to be 2.72 m³ per year (table 8). The Muslims used less wood (including charcoal) for fuel than the Buddhists, the average amounts being about 2.10 and 3.33 m³ per household per year respectively. The average household consumption of firewood in Ko Lao was

FIG. 6.   Heavily used mangrove forest near Had Sai Khao

FIG. 7.   Cooking is a major use of mangrove wood in Ko Lao

TABLE 8. Estimated consumption of fuelwood per household by residents of Ko Lao and Had Sai Khao

	Firewood			Charcoal				Total
	sticks per day	sticks per year	m³ per year	kg per day	kg per year	m³ per year	m³ wood per year	fuelwood (m³ per year)
Ko Lao								
Buddhists	3.8	1,388	2.21	0.96	352.2	0.67	1.12	3.33
Muslims	4.1	1,509	1.28	0.90	330.0	0.49	0.82	2.10
average	3.9	1,448	1.75	0.93	341.0	0.58	0.97	2.72
Had Sai Khao	0.6	200	0.32	1.9	680	1.0	1.7	2.02

about 1,448 sticks, or 1.8 m³, per year. (The firewood sticks used by the Buddhist villagers are about 40 cm long and 6 cm in diameter, whereas the Muslims use a smaller size, 4 cm in diameter.) Charcoal consumption per household was 0.58 m³, or 341 kg, per year. All the charcoal used in the village is bought in Ranong or in neighbouring villages, especially Had Sai Khao, where it is produced.

Most Had Sai Khao villagers used locally produced charcoal for cooking. The charcoal is made in small masonry or earth-pit kilns (figs. 8–10). The average consumption of charcoal per household was 1 m³, or 680 kg, per year — about twice the amount used in Ko Lao (table 8). Most of the firewood used in Had Sai Khao is burned in order to get rid of mosquitos and other insects; consumption was only about 200 sticks, or 0.3 m³, per household per year. The average total fuelwood consumption per household in Had Sai Khao was about 2.02 m³ per year.

Fishing Gear

Some of the several types of fishing gear used by Ko Lao and Had Sai Khao villagers are made of mangrove wood. Poles from *Rhizophora* and *Bruguiera* spp. are used for crab traps. The amount of wood used for this equipment is minimal, because only small poles, about 3–5 cm in diameter and 3–4 m long, are used (see fig. 16). About 60 per cent of Ko Lao households use crab traps, with about 30 traps per household. About 70 per cent of Had Sai Khao households use crab traps, with each household having 30–50 traps. The quantity of mangrove wood used for crab traps was estimated to be 0.10–0.3 m³ per household. Posts made from *Rhizophora apiculata* and *Bruguiera* spp. are also used for drift gill nets. One type of fishing equipment that uses a lot of mangrove poles is the winged set bag, but this

type of equipment was rarely seen in these two villages. It can be concluded that the volume of wood used for fishing equipment in both villages is small.

Housing

The typical houses of Ko Lao and Had Sai Khao villagers are very similar (fig. 11). Most houses have a bedroom, a living room, and a kitchen, with platforms in front of and behind the house. The platform behind the house is generally used for drying prawns, fish, and shrimp paste. Roofs are usually made of nipa leaves, walls of bamboo, floors and platforms of mangrove wood. A few houses in Ko Lao and Had Sai Khao have floors and/or walls and/or platforms made of wooden planks, and roofs of zinc-coated corrugated iron. Mangrove species commonly used for house construction are shown in table 9.

One Ko Lao house was measured to estimate the quantity of wood used in construction. This house was 4 x 4 m and comprised a bedroom, a living

TABLE 9. Common mangrove species used for house construction in mangrove villages

Columns (posts)	*Rhizophora apiculata*
	R. mucronata
	Avicennia alba
	Bruguiera parviflora
	B. cylindrica
	Xylocarpus spp.
Bracing members	*Rhizophora* spp.
	Bruguiera spp.
Beams (girders) and subflooring	*Rhizophora* spp.
	Xylocarpus spp.
Roof frame, partition frame, floor, and platform (wooden terrace)	*R. apiculata*
	Bruguiera spp.

FIG. 8. *Rhizophora* wood loaded into a small earth-pit kiln for making charcoal at
Had Sai Khao

FIG. 9. Firing a small earth-covered charcoal pit at Had Sai Khao

FIG. 10.   Unloading the charcoal kiln after burning for four or five days

FIG. 11.   House construction using mangrove lumber, Ko Lao

room, and a kitchen, with platforms in front (2 x 4 m) and behind (3 x 6 m). The roof was made of nipa leaves, and the walls were bamboo. Two Had Sai Khao houses were also measured. The first house was rather big, about 8 x 6 m, and comprised a bedroom, a big living room, and a kitchen, with a platform in front (4 x 8 m). The roof was made of corrugated iron, and the floor and walls were of planks. The other house was smaller, about 6 x 5 m, comprising a bedroom, a living room, and a kitchen, with a front platform (6 x 4 m). The roof was made of nipa leaves, and the walls were bamboo. The total amount of wood used to construct the house in Ko Lao was estimated at about 9 m^3, the larger Had Sai Khao house 19 m^3, and the smaller Had Sai Khao house 11 m^3 (table 10).

Villagers' Perceptions of Mangrove Resources

Villagers living in mangrove areas use mangrove trees for their basic subsistence in the form of fuelwood, for construction of houses and fishing gear, and for catching marine animals. All the Had Sai Khao residents and most families in Ko Lao (70–80 per cent) viewed the mangrove forest primarily as the source of forestry and fishery products for daily life. About 20–30 per cent of Ko Lao households expressed no opinion on this topic.

Long-term residents, who had lived in Ko Lao for about 10 years and in Had Sai Khao for almost 20 years, said that more of the forest is being destroyed every year. About 65 per cent of the mangrove villagers in Ko Lao and 80 per cent in Had Sai Khao believed that the main cause of destruction is the cutting of trees by concessionaires, whom the villagers called "the owners of the charcoal kilns."

Residents of both villages believed that the population of marine animals declined after mangrove forests were destroyed. They also said that their daily catch of fish, shrimp, and crabs is less now than 10 or 20 years ago. They believed this

TABLE 10. Estimated amount of wood used in the construction of three houses in Ko Lao and Had Sai Khao

	Diameter (cm)	Length (m)	Number of pieces	Estimated volume (m^3)		
				4 × 4m^2 house	6 × 5m^2 house	8 × 6m^2 house
Posts	12–18	8–10	16	—	—	4.1
	8–12	6–8	35	—	—	3.2
	10–16	8–10	16	—	3.2	—
	8–10	6–8	18	—	1.1	—
	10–16	8–10	18	3.6	—	—
Bracing members	6–8	6–8	10	0.4	—	—
Beams	10–16	8	12	—	—	1.9
	10–16	7	10	—	1.4	—
	10–16	6	8	1.0	—	—
Subflooring	8–10	8	36	—	—	2.7
	6–8	7	36	—	1.3	—
	6–8	6	36	1.1	—	—
Roof frame	6–8	8	30	—	—	1.2
	6–8	6–8	26	—	1.0	—
	4–6	6	20	0.4	—	—
Wall frame	4–6	6–8	30	—	0.7	—
	4–6	6	24	0.4	—	—
Floor	6–8	6–8	90	—	—	3.6
	4–6	6	62	—	1.1	—
	4–6	4–5	58	0.8	—	—
Platform	6–8	4–6	86	—	—	2.6
	4–6	4–6	68	—	1.2	—
	4–6	4–6	60	1.0	—	—
Total volume				8.7	11.0	19.3

resulted from the destruction of the mangrove forest and from the increase in the number of people who had come to live in and use the mangrove area.

Ko Lao and Had Sai Khao villagers complained that most of the benefit from the use of mangrove forests goes to the concessionaires. They would like to see the mangrove forest protected mainly for fishery production to assure continuation of their daily subsistence. Mangrove residents, particularly those in Had Sai Khao village, wanted the Royal Forest Department to allow them to cut at least enough mangrove trees for fuel, construction of houses, and fishing equipment.

Villagers believed that the concessionaires, who obtain the greatest benefit from the forest, should take responsibility for protecting the forest from illegal cutting, maintaining the balance of the ecological system and sustaining both forestry and fishery resources. Mangrove villagers would like to take over the responsibility for protection of forests around the villages, if the concessionaires would hire them. They also wanted to work in such activities as planting and logging, in order to earn cash income for their daily subsistence, but the concessionaires have never hired them. Concessionaires usually bring in their own labourers from elsewhere.

Few Ko Lao villagers indicated that they knew about the roles of concessionaires and government foresters in the management and use of mangrove resources. The level of communication between the Ko Lao villagers and the concessionaires is very low. Had Sai Khao villagers have more contact with the concessionaires and government foresters. The village headman or some other villagers used to report to the chief of the Mangrove Management Unit and the research scientists at the Mangrove Research Station or to the concessionaires when they saw people from other villages cutting mangrove trees illegally.

The perceptions of Ko Lao and Had Sai Khao villagers concerning most aspects of the mangrove forests were very similar except for their beliefs about protection of the forest. Most of the people in Had Sai Khao believed in forest spirits. They had built a small house for the mangrove-forest spirit (fig. 13) and made offerings to it. They believed that the spirit was effective in preventing anyone from destroying the forest and had caused the forest near Had Sai Khao to be more fertile than in other parts of this area. No spirit house was seen in Ko Lao.

The Buddhists at Ko Lao village seemed to have more understanding of the ecology of mangrove forests and the processes of mangrove manage-

FIG. 12.   Boat building using mangrove lumber, Ko Lao

ment and utilization than the Muslims. One probable reason is that the Buddhists in general were better educated and also had more opportunity to communicate with people whose work is related to mangrove forests, such as forest officers from the Mangrove Management Unit, concessionaires, and workers at the charcoal kilns. The language barrier was also a problem for some of the Muslim people, but it was not serious because only a few of them (mostly older men) could not speak Thai. In addition, the Muslims usually worked with other Muslims because they wanted to avoid quarrels with Buddhists living in the same village or in other villages. It was learned that they once had a quarrel with Buddhists from other villages about cutting trees and catching marine animals.

Fishery

Research Orientation

Fishing methods and the species caught were studied at Ko Lao and Had Sai Khao through observation, participation in fishery activities, and interviews. The study was aimed at improving the life of small-scale fisherfolk in these two villages by increasing their daily earnings.

Ko Lao and Had Sai Khao have relatively rich natural resources for fishing and for gathering mangrove forest products. An appreciation of their ethic of self-sufficiency and of the working symbiosis between the people and their environment (where preservation of biological diversity is essential for maintaining options for future generations) was a prerequisite for this investigation. The programme studied the basic needs of the inhabitants of the villages: food, safe water, clothing, shelter, health, and education, consistent with the goals of the Fifth National Socio-economic and Development Plan.

Capture Fishery

The methods used in capture fishery in different maritime provinces vary, but some common fishing methods were used by the villagers of Ko Lao and Had Sai Khao. In determining the efficiency of capture fisheries, one must take into account the kinds of fishing gear and the total catch. Fishing is the main occupation in both Ko Lao and Had Sai Khao. Because Ko Lao households generally used only small to medium-sized fishing boats, their income was generally lower than that of Had Sai Khao households. The people live at the edge of the land, close to the mangrove forest. Because

their boats are small, their fishing is confined to the inshore and near-shore zones. June to October is the best period for fishing these two zones, and so income is not spread evenly through the year.

Fishing Gear

Only a few of the types of fishing operations traditionally practised in Thailand are carried out along the shores of Ko Lao, Had Sai Khao, and adjacent estuaries. The gear varies in structure and size, as described below.

Scissor nets. In Ko Lao mysids and other small crustaceans such as a species of *Acetes* are caught extensively on the shallow sandy mud flats with brine-shrimp scissor nets. The scissor net consists of a fine-mesh bag net mounted on two long, slender bamboo poles. Wooden shoes are attached to the lower ends of the poles to prevent them from sticking in the mud (see fig. 14). Small scissor nets are operated by hand, but the larger ones are generally operated from motorized boats. The small crustaceans are harvested from the sandy mud at depths of 30 to 120 cm. They seem to be abundant during the rainy season, from May to August. They occasionally disappear when scissor nets of a larger type used to catch fish and penaeid shrimp are employed in the shallow waters, causing high turbidity. The brine-shrimp scissor net is used only by Ko Lao fishers.

Push nets. About 28 per cent of the total shrimp catch (excluding mysids) was taken in push nets, consisting of a bag of fine mesh (approximately 0.8–1.8 cm) mounted on two sturdy bamboo poles about 5 to 8 m long. This gear is operated from a motorized boat of 60 to 220 hp, by two workers. The catch also includes young and juvenile fish, crabs, and molluscs. This type of net is commonly used by the people of Had Sai Khao village.

Barrier nets. Often the mouth of a creek or channel in the mangrove forest is closed with a barrier net, consisting of several nets connected, up to about 100 m in length. The lower side of the net is attached to the mud flat at low tide with small pieces of bamboo. Each end of the net is mounted on a large, strong bamboo pole so that it can be stretched across the waterway. At high tide the fishermen lift the upper side of the net to the surface of the water to trap fish when the tide ebbs. The size of fish caught depends on the size of the mesh, which is about 2.5 to 5 cm. Fish of economically important species from the catch may

FIG. 13   House for the mangrove-forest spirit, near Had Sai Khao

FIG. 14.   Drying shrimp and fish on a house platform, Ko Lao. The wooden shoes of a brine-shrimp scissor net are to the right of the boy; a cage for keeping live fish is to the left.

be kept alive in bamboo cages in order to get a higher price.

Crab net traps. Baited traps are used by the people of both villages to catch mangrove crabs (*Scylla serrata*). The crabs are caught mainly for houshold consumption. The simple gear consists of a ring of bamboo about 45 cm in diameter mounted about 30 cm above the lower, pointed end of a thin mangrove pole, with a coarse net stretched in the ring (fig. 16). Shark meat is normally fixed to the centre of the ring as bait. The traps are driven into the mud close to the mangroves. They are placed from a row boat at high tide and are pulled up after three or four hours. The clumsy crabs cling to the net when it is lifted out of the water.

Gill nets. Gill nets are not commonly used by the people of Ko Lao, because the coastal waters are shallow, but they are used along the coast of Had Sai Khao, where the water is deeper. The net is usually more than 100 m long, with a 7.5 to 10 cm mesh, and is operated from the shore. The catch consists mostly of pelagic fish. Swimming crabs (*Portunus pelagicus*) are often caught along with the fish.

Bag nets. Winged bag nets mounted across mangrove channels are found in both Ko Lao and Had

Sai Khao and in adjacent areas, though this type of gear is not very popular because it cannot be moved. The net is lowered during high tide to trap the fish, which are harvested just before low tide. Many varieties of fish, shellfish, and other marine animals are captured.

Other fishing operations. Several minor fishing operations that are occasionally used include scoop-netting for jellyfish, gathering molluscs, and using single hooks, long lines, off-shore traps, and cast nets.

Jellyfish (coelenterates) are scooped from the surface of the water from motorized boats between October and December, using simple round scoop nets (fig. 17). The jellyfish are salted and dried before being sold to a middleman.

Some families in both villages gather *Modiola, Anadara, Perna,* limpets, and other edible molluscs for home consumption.

Only a few families from the two villages fish with single hooks or long lines. Long lines of about 50 to 100 m (fig. 18) are operated close to the mangrove edge, while single hooks are dangled from twigs or branches of mangrove trees. The catch usually consists of young grouper (*Epinephalus*

FIG. 15.   Fisherman with his small boat engine

FIG. 16.   Mangrove poles serve as handles for round-net crab traps

FIG. 17.   Scoop nets for jellyfish

FIG. 18.   Long-line fisherman, Had Sai Khao

FIG. 19.   Fish trap

tauvina), sea bass (*Lates calcarifer*), and other non-economic fish.

One household of Ko Lao sometimes uses off-shore traps for catching coral-reef fish such as *Siganus* and *Epinephalus* (fig. 19).

Only one household in Had Sai Khao uses a cast net. Shrimp is the major catch, but the catch from each operation is very small.

Composition of the Catch

As mentioned earlier, these traditional fishing methods are confined to the inshore and near-shore zones, so only juvenile or young stages of economic species are captured. The catch at Ko Lao and Had Sai Khao includes three major categories: fin fish, crustaceans, and molluscs. The species most commonly taken are similar in the two communities and are listed in table 11. The catch of penaeid shrimp is very small in Ko Lao, but because of the use of push nets in Had Sai Khao, shrimp are a major catch there.

Total Value of the Catch

In Ko Lao, fish are sun-dried and salted for sale, while Had Sai Khao people more commonly sell fresh fish (figs. 20 and 21).

Ko Lao. Fishing was the main occupation in 21 of the 26 households in Ko Lao (table 12). There were 11 households (42 per cent) that caught fish and small crustaceans; 8 (31 per cent) caught only small crustaceans; and 2 (8 per cent) harvested molluscs in addition to catching fish and small crustaceans. The other 5 households (19 per cent) probably fished only occasionally.

Small crustaceans, including mysids and *Acetes,* are plentiful in the water around Ko Lao. There were 24 households that gathered these crustaceans. Each household's total catch depends on the number of workers in the family, the number and size of nets used, and the duration of fishing. The annual catch for the different households varied from 270 to 2,000 kg, giving a total of 22,365 kg per year for the 24 operating households (or a mean of 932 kg per household). These small crustaceans are sold fresh or as shrimp paste, which is an important ingredient in South-East Asian cooking. To make the paste, the crustaceans are half-dried, ground with salt, then fermented, and dried again (fig. 22). Two kilo-grams of fresh crustaceans yield 1 kg of shrimp

paste. The value of the catch, in terms of shrimp paste after processing, was about 456,006 baht (a mean of B 19,000 per household).*

There were 16 households in Ko Lao that engaged in catching fish — using long lines, gill nets, barrier nets, hooks, set bag nets, and other small gear. The fishing areas within one or two kilometres of Ko Lao are sand flats exposed during low tide. The total catch was about 11,280 kg, valued at B 215,164 per year (means of 705 kg and B 13,448 per household per year). Sale prices for the catch vary according to species of fish, size, and season.

Three households gathered crabs, using either

*B 23 = US$1 on 31 December 1983.

TABLE 11. Species of marine animals commonly caught by Ko Lao and Had Sai Khao fishers

	Local name	Scientific name
Fish	kod ta-le	*Tachysurus connulata*
	krabok	*Mugil dussumeri*
	kao	*Epinephalus tauvina*
	krud-krad	*Pomadasys* sp.
	kang-lai	*Therapon jabua*
	koak	*Anodontostoma* sp.
	juad	*Otholithes* sp.
	haang-kwai	*Thysanophrys* sp.
	krapong-khangpaan	*Lutianus* sp.
	krapong	*Lates calcarifer*
	salid-hin	*Siganus javers*
	krabane	*Dasyatis* spp.
	takrub	*Scathophagus argus*
	duk ta-le	*Plotosus lineatus*
	noalchan	*Chanos chanos*
	meow	*Engraulis* spp.
	kem ta-le	*Hemirhamphus* sp.
Crustaceans	kung khao	*Penaeus merguiensis*
	kung kula-dum	*Penaeus monodon*
	kung keoy	*Acetes* sp.
	poo ta-le	*Scylla serrata*
	poo ma	*Portunus pelagicus*
	kang tak-taan	*Squilla mantis*
Molluscs	hoi malang-poo	*Perna viridis*
	hoi kra-pong	*Modiola* sp.
	hoi krang	*Anadara* sp.
	hoi paak-pet	[limpets]
	hoi mhuk klouy	*Loligo* sp.
	hoi mhuk kradong	*Sepia* sp.

FIG. 20.   Salted fish drying for home consumption

FIG. 21.   Fish and shellfish caught by Had Sai Khao fishers

crab traps, set bag nets, or gill nets. *Scylla serrata* and *Portunus pelagicus* are the economically important crabs of the area. They are usually caught close to the mangrove edge. The total catch was approximately 860 kg per year, valued at B 19,125 (means of 286.7 kg and B 6,375 per household per year).

Several families looked for additional sources of food for family consumption, including gathering molluscs of species commonly found on the mudflats or diving for *Modiola, Perna,* and *Anadara.*

The total value of the economic species caught in Ko Lao was estimated at B 690,295 (B 456,006 for shrimp in the form of paste, B 215,164 for fish, and B 19,125 for crabs), giving an average income for the 26 households of this traditional fishing community of about B 26,550 per year, or B 2,212 per month.

TABLE 12.   Households in Ko Lao catching various categories of aquatic animals

| | Number of households | | | |
	Buddhist	Muslim	Total	%
Small crustaceans and fish	9	2	11	42.3
Small crustaceans only	3	5	8	30.8
Small crustaceans, fish, and molluscs	—	2	2	7.7
Small crustaceans and molluscs	—	1	1	3.8
Small crustaceans and crabs	—	1	1	3.8
Small crustaceans, fish, and crabs	1	—	1	3.8
Fish only	—	1	1	3.8
Fish, penaeid shrimp, and crabs	1	—	1	3.8
Total	14	12	26	99.8

FIG. 22.   Small crustaceans (mysids or *Acetes* sp.) in the half-dry stage, for making shrimp paste

TABLE 13.   Households in Had Sai Khao catching various categories of aquatic animals

	Number of households	%
Fish, crabs, and molluscs	1	3.8
Fish, crabs, and shrimp	5	19.2
Fish and crabs	10	38.6
Fish and shrimp	3	11.5
Crabs and shrimp	2	7.7
Fish only	1	3.8
Crabs only	4	15.4
Total	26	100.0

Had Sai Khao. All 26 households in Had Sai Khao were engaged in catching fish, shrimp, molluscs, or crabs. There were 22 households that fished for crabs, 20 for fish, and 10 for shrimp (table 13).

The average catch per household per year was about 925 kg of fish, 452 kg of shrimp, 662 kg of crabs, and 4 kg of molluscs (table 14). Part of the catch was consumed at home and the rest was sold to the market (however, all molluscs gathered were consumed at home). Size influences the selling price. Prices for crabs vary less than those for fish or shrimp because the bigger crabs are usually selected for sale. The total annual sale of fish, shrimp, and crabs from the village of Had Sai Khao came to B 1,423,318. This yielded a moderately high average income per household of B 54,743 per year, or B 4,562 per month — derived more from shrimp than from fish or crabs (table 14).

Culture Fishery: Shrimp Farming

One attempt was made to culture shrimp at Ko Lao some years ago. The shrimp farm, which was located only five or ten metres inside the mangrove forest, was not successful because of poor management and location. The owner believed that he only had to construct the ponds and pump in tidal waters carrying shrimp seed, and that after one or two months the shrimp could be harvested. Shrimp-farm management is more complex. The location of a shrimp pond must be carefully selected, and physico-chemical and biological factors need to be studied first. Primary productivity may be possible where an abundance of shrimp seed exist in coastal waters of high quality, but in ponds constructed in mangrove forest the texture of the soil is too loose for dike construction and cannot withstand the pressure of tidal water (fig. 23).

Socio-economic Conditions

General Features

Population origins. The present settlements of Ko Lao and Had Sai Khao were founded more than 23 years ago. Long before that the two village sites had been inhabited by a Muslim group of unknown origin, called *chao nam* in Thai, meaning fishermen. None of the people in the two villages seemed to know where this pioneer group came from or where they had gone.

The present population of Ko Lao moved there from the south, while the majority of the Had Sai Khao population came from central Thailand. This accounts for the different dialects spoken in the two villages; Ko Lao dwellers use a southern dialect or Malay, and those in Had Sai Khao speak Central Thai.

Reasons for migration to mangrove villages. The most common reason given by household heads in both Ko Lao and Had Sai Khao for migrating to the area was the attraction of the abundance of

TABLE 14.   Average annual catch and income per household from aquatic animals — Had Sai Khao

	Catch (kg)	Consumption (kg)		Sale price (baht/kg)		Income (baht)
		Home	Market	Average	Range	
Fish	924.8	92.1	832.7	30	20–50	18,078
Shrimp	452.3	72.7	379.6	70	40–80	24,105
Crabs	661.7	32.8	628.9	19	18–30	12,560
Molluscs	3.9	3.9	—	—	—	—
Total	2,042.7	201.5	1,841.2			54,743

FIG. 23.   Collapsed sluice gate of a failed shrimp pond at Ko Lao

marine life and mangrove forest. Many were facing problems of population pressure and had less opportunity to earn an adequate income where they came from. They were looking for areas like Ko Lao and Had Sai Khao where there were still plenty of marine animals to catch.

Among those not born in Ko Lao village, almost 91 per cent of the Buddhist respondents were cash-crop farmers before they moved to Ko Lao; none of them had been fishers. For the Muslims the previous occupation was predominantly fishing; about 29 per cent had been wage labourers, and 11 per cent were formerly orchard owners.

Reasons for migrating to Had Sai Khao are implied by the former occupations of the household heads. Prior to moving to Had Sai Khao, almost 62 per cent of the respondents had faced a shortage of farm land for growing rubber or other cash crops. About 38 per cent of respondents said they had moved to Had Sai Khao because they wanted to pursue a career in fishing.

Household size and composition. The Ko Lao households were larger on average (5.8 persons) than the Had Sai Khao households (4.6 persons) (table 15) and had a higher proportion of children (table 16), suggesting that Ko Lao families have a higher birth rate.

There was an average of 2.8 persons of working age (15–54 years) in each Ko Lao household, or about 48 per cent of the household members (table 15). The dependency ratio — i.e. the ratio of dependants (mostly children under age 15) to family members of working age — was about 1.07, meaning that each worker has to support one dependant in addition to him- or herself. The ratio was much higher in Muslim households (1.28) than in Buddhist households (0.86), principally because of the higher proportion of children in Muslim households.

In Had Sai Khao there were about 2.7 persons of working age per household, or 58 per cent of the population, and a dependency ratio of 0.72. The difference in dependency ratios suggests that Ko Lao households, especially the Muslim households, had a heavier burden of looking after their dependants than had the Had Sai Khao households.

Education. In order to increase the literacy of rural people, Thailand has for a long time had legislation requiring all children to enter primary school at age 7, and school attendance is now compulsory between the ages of 7 and 13, corresponding to the six primary grades. Despite this requirement, 23 per cent of the total population was illiterate in Ko Lao, and 17 per cent in Had Sai Khao (table 17). Fortunately, educational opportunities are better

TABLE 15. Average numbers of all members and of members of working age in households in Ko Lao and Had Sai Khao, by sex and religion

	Number of households	Number of persons	Persons per households	Workers per households[a]
Ko Lao				
Buddhist households	14			
males		41	2.9	1.4
females		37	2.6	1.6
total		78	5.6	3.0
Muslim households	12			
males		36	3.0	1.3
females		37	3.1	1.3
total		73	6.1	2.7
All households	26			
males		77	3.0	1.4
females		74	2.8	1.5
total		151	5.8	2.8
Had Sai Khao				
All households	26			
males		58	2.2	1.3
females		61	2.3	1.3
total		119	4.6	2.7

Slight discrepancies in the totals are due to rounding.
a. Household members between the ages of 15 and 54 years.

TABLE 16. Composition (percentages) of the population by age, sex, and religion

Age (years)	Ko Lao									Had Sai Khao		
	Buddhists			Muslims			Total					
	M	F	T	M	F	T	M	F	T	M	F	T
0–4	10.3	5.1	15.4	6.8	11.0	17.8	8.6	7.9	16.6	5.0	3.4	8.4
5–9	5.1	7.1	12.8	6.8	6.8	13.7	6.0	7.3	13.2	4.2	3.4	7.6
10–14	7.7	3.8	11.5	6.8	8.2	15.0	7.3	6.0	13.2	8.4	10.1	18.5
15–19	5.1	9.0	14.1	5.5	9.6	15.0	5.3	9.3	14.6	6.7	10.1	16.8
20–24	5.1	2.6	7.7	6.8	2.7	9.6	6.0	2.6	8.6	4.2	3.4	7.6
25–29	2.6	1.3	3.8	4.1	4.1	8.2	3.3	2.6	6.0	4.2	2.5	6.7
30–34	3.8	3.8	7.7	1.4	1.4	2.7	2.6	2.6	5.3	1.7	1.7	3.4
35–39	3.8	2.6	6.4	1.4	1.4	2.7	2.6	2.0	4.6	2.5	3.4	5.9
40–44	0	1.3	1.3	2.7	1.4	4.1	1.3	1.3	2.6	3.4	3.4	6.7
45–49	3.8	1.3	5.1	0	0	0	2.0	0.7	2.6	3.4	1.7	5.0
50–54	1.3	6.4	7.7	0	1.4	1.4	0.7	4.0	4.6	2.5	3.4	5.9
55–59	1.3	1.3	2.6	5.5	1.4	6.8	3.3	1.3	4.6	0.8	1.7	2.5
60–64	2.6	0	2.6	0	0	0	1.3	0	1.3	0.8	0.8	1.7
65–69	0	0	0	1.4	1.4	2.7	0.7	0.7	1.3	0	0.8	0.8
70–74	0	1.3	1.3	0	0	0	0	0.7	0.7	0.8	1.7	2.5
75+	0	0	0	0	0	0	0	0	0	0	0	0
Total	52.6	47.5	100	49.3	50.7	100	51.0	49.0	100	48.7	51.3	100
Number of persons	41	37	78	36	37	73	77	74	151	58	61	119

Slight discrepancies in the totals are due to rounding.

TABLE 17. Educational status

| | Ko Lao | | | | | | Had Sai Khao | |
| | Buddhists | | Muslims | | Total | | | |
	No.	%	No.	%	No.	%	No.	%
Illiterate	14	17.9	20	27.4	34	22.5	20	16.8
Below school age[a]	18	23.1	21	28.8	39	25.8	17	14.3
Students	14	17.9	14	19.2	28	18.5	23	19.3
Schooling completed								
<4 years	8	10.3	2	2.7	10	6.6	18	15.1
4 years	16	20.5	14	19.2	30	19.9	28	23.5
5–7 years	7	9.0	2	2.7	9	6.0	11	9.2
high school	1	1.3	0	0	1	0.7	2	1.7
Total	78	100	73	100	151	100	119	100

a. Children less than seven years old are not required to attend school.

than in the past. There is now a primary school with six teachers in Ko Lao and one with five teachers in Had Sai Khao. Both schools offer education through grade 6. A relatively large proportion of the residents of both villages had completed grade 4 (20 per cent in Ko Lao, 24 per cent in Had Sai Khao) and thus had acquired basic skills of literacy.

Land ownership. All the land on Ko Lao and Had Sai Khao islands has been declared public land. Because the mangrove forests are part of the forest reserve, the mangrove dwellers cannot own land in the villages, but they are allowed to build houses on the seashore around the islands. The houses are built close together, each home lot averaging only about 15 to 20 m².

Village leadership. There was no formal organized group for community development in either village. The village headman appeared to receive the highest credit from all respondents in each community as the leader of a community development project, the next most important leader being the teacher.

Common problems of life. An adequate supply of water for drinking and other domestic uses was a common problem for the two mangrove village (see ch. 3, below) (fig. 24). Another problem reported by households in both villages was sudden illness. The overall level of well-being, health care, and sanitation in the villages was very poor. They urgently need a programme to solve these problems, especially those concerning young people, to provide a better quality of life in the future.

Household Economy

Although Ko Lao and Had Sai Khao are located close to each other on separate islands, there are many economic differences between them. On the average, Ko Lao households were poorer than Had Sai Khao households. Shrimp-paste making was the major source of income for Ko Lao dwellers, whereas sea fishery without processing generated the major income for Had Sai Khao villagers. No Ko Lao households earned income from making charcoal, but two Had Sai Khao households obtained supplemental income by making and selling charcoal. No other income was earned from forestry sources by mangrove dwellers of either village.

Table 18 shows that the average annual income earned by Ko Lao households was about B 31,885. The largest contribution to household income in Ko Lao was from shrimp paste, which generated B 17,539 per household, or 54 per cent of the total. Fishing contributed B 9,011 per household, almost 30 per cent. Wage work accounted for only 12 per cent, but was especially important for those households that had a large labour force. On the average, the Buddhist households earned more annual income (B 36,700) than the Muslim households (B 26,268). Because of the difference in household size, per capita income in Buddhist households (B 6,673) was about 55 per cent higher than that in Muslim households (B 4,306).

FIG. 24.   One of the three shallow wells at Ko Lao

TABLE 18.   Annual income per household from all sources

| | Ko Lao | | | | | | Had Sai Khao | |
| | Buddhists | | Muslims | | Total | | | |
	Baht	%	Baht	%	Baht	%	Baht	%
Unprocessed fish	10,174.5	27.7	7,653.8	29.1	9,011.1	28.3	54,743.3	85.6
Shrimp paste	19,075.6	52.0	15,745.6	59.9	17,538.7	54.4	0	0
Charcoal	0	0	0	0	0	0	252.2	0.4
Wage labour	6,021.4	16.4	1,118.3	4.3	3,758.4	11.8	5,050.1	7.9
Grocery shop	1,142.9	3.1	1,200.0	4.6	1,169.3	3.7	2,010.2	3.1
Trader/fish merchant	0	0	0	0	0	0	1,923.1	3.0
Other	285.5	0.8	550.0	2.1	407.6	1.3	0	0
Total	36,699.9	100	26,267.7	100	31,885.1	100	63,978.9	100

The average annual income in Had Sai Khao was B 63,979 pr household, or more than twice that of Ko Lao households (table 18). In contrast to Ko Lao, there was no shrimp-paste making in Had Sai Khao, and most of the villagers had never been involved in making shrimp paste. Fishing without processing contributed nearly 86 per cent to the total household income. Wages were the next most important source of income for Had Sai Khao families, accounting for about 8 per cent of the total. The demand for hired labour, including fishing crews, came from households fishing on a relatively large scale. Thus households in Had Sai

Khao with greater numbers of adult males had greater opportunity to hire out family members.

Two Had Sai Khao households made charcoal commercially, while four households produced charcoal for their own use. Had Sai Khao fishermen had the option of selling their catch either to village merchants or to merchants in Ranong. Two households bought fish for resale in Ranong and also engaged in fishing for themselves, but most households sold their catch to the two merchants in their village since it was more convenient. They might also sell to merchants in Ranong when their catch was large enough to cover the shipping costs from the village.

The means, ranges, and standard deviations of household cash income from various sources in the two villages are given in table 19. These figures suggest the range of variation in the sources and amounts of household income in the villages.

Production and consumption expenses. A summary of the average production expenses incurred by households in the two communities in fishing and in making shrimp paste and charcoal is given in table 20. The average annual production costs of Ko Lao households were B 4,400 for fishing and B 11,996 for making shrimp paste, totalling

an average of B 16,396 per household per year. These costs accounted for 51 per cent of the total household income. It should be pointed out that these costs include both cash and non-cash or unpaid expenses, such as family labour and household production of small shrimp. Expenses for such items as salt, fuel, lubricants, repairs, and transportation accounted for only 14 per cent of the total production cost of shrimp paste.

Food accounted for 84 per cent of all household consumption expenses in Ko Lao and 83 per cent in Had Sai Khao (table 21). Total consumption and production costs exceeded the annual income of Ko Lao households. They could survive economically because 50 per cent of the value of their food consumption (other than rice) was not paid

TABLE 19. Means, ranges, and standard deviations (SD) of household income (baht per year), by source of income

	Ko Lao			
	Buddhists	Muslims	Total	Had Sai Khao
Unprocessed fish				
mean	10,174.5	7,653.8	9,011.1	54,743.3
range	0–103,533	0–75,366	0–103,533	5,869–204,921
SD	19,788.6	19,582.2	19,733.4	54,479.5
Shrimp paste				
mean	19,075.6	15,745.6	17,538.7	0
range	0–67,488	0–44,268	0–67,488	
SD	7,354.0	6,751.7	7,274.3	
Charcoal				
mean	0	0	0	252.2
range				0–3,934
SD				892.3
Wage labour				
mean	6,027.4	1,118.3	3,758.4	5,050.1
range	0–19,544	0–15,635	0–19,544	0–26,200
SD	6,457.8	5,627.7	6,136.7	7,718.5
Grocery shop				
mean	1,142.9	1,200.0	1,169.3	2,010.2
range	0–16,000	0–14,000	0–16,000	0–36,533
SD	4,276.2	3,979.9	4,056.4	7,537.3
Trader/fish merchant				
mean	0	0	0	1,923.1
range				0–50,000
SD				6,697.7
Other				
mean	285.5	550.0	407.6	0
range	0–492	0–2,952	0–2,952	
SD	246.9	1,092.9	822.2	
Total income				
mean	36,699.9	26,267.7	31,885.1	63,978.9
range	10,003.8–119,533.7	5,625.7–99,589.7	5,625.7–119,533.7	6,017.8–131,114.2
SD	29,887.1	25,927.4	31,015.3	52,005.6

TABLE 20.   Average production expenses per household for fishing, making shrimp paste, and making charcoal

| | Ko Lao | | | | | | Had Sai Khao | |
| | Buddhists | | Muslims | | Total | | | |
	Baht	%	Baht	%	Baht	%	Baht	%
Fishing	6,082.2	31.8	2,437.4	18.5	4,400.0	26.8	19,242.3	99.3
Shrimp paste	13,047.7	68.2	10,769.9	81.5	11,996.4	73.2	0	0
Charcoal	0	0	0	0	0	0	126.1	0.7
Total	19,129.9	100	13,207.3	100	16,396.4	100	19,368.4	100

Includes estimated cash value of labour and other non-cash inputs to production.

TABLE 21.   Average annual consumption per household

| | Ko Lao | | | | | | Had Sai Khao | |
| | Buddhists | | Muslims | | Total | | | |
	Baht	%	Baht	%	Baht	%	Baht	%
Food								
rice	9,877	35.2	5,395	24.3	7,808	30.8	5,994	13.8
other food	13,861	49.4	12,913	58.2	13,424	52.9	30,078	69.0
subtotal	23,738	84.5	18,308	82.5	21,232	83.7	36,072	82.8
Non-food expenditures								
clothing	1,033	3.7	993	4.5	1,015	4.0	2,185	5.0
medical	1,181	4.2	821	3.7	1,015	4.0	1,735	4.0
education	208	0.7	88	0.4	152	0.6	171	0.4
charity	135	0.5	540	2.4	322	1.3	447	1.0
entertainment	757	2.7	388	1.7	587	2.3	755	1.7
housing maintenance	333	1.2	400	1.8	364	1.4	533	1.2
charcoal	701	2.5	657	3.0	681	2.7	1,226	2.8
other	0	0	0	0	0	0	467	1.1
subtotal	4,348	15.5	3,887	17.5	4,135	16.3	7,519	17.2
Total	28,086	100	22,195	100	25,368	100	43,591	100

Includes estimated cash value of household products consumed at home.

for in cash but was drawn in kind from family labour or household production.

Net family income. Net family income is defined as the difference between total household income and total household expenses of production and consumption. Table 22 shows the average net household income of the two surveyed villages, calculated by subtracting all cash and non-cash expenses from the total household cash income. These figures show clearly the importance of non-cash transactions in the household economy of these mangrove villagers. If all transactions were

based on cash, the households of Ko Lao village would have had a negative average annual income of B 9,879 per household. The appearance of a negative net income is possible because non-cash or unpaid expenses are included in the estimates of production costs and consumption. Total cash expenses can be derived if all unpaid items are excluded from production and consumption expenses. These results are also shown in table 22. Ko Lao households spent an average of B 25,832 in cash for their family production and consumption. When all cash expenses are subtracted from the household income, the net family cash income averaged B 6,053 per year, or B 504 per month. Buddhist households had a higher net cash income (B 7,497) than Muslim households (B 4,368).

Had Sai Khao households had a higher net total income and a higher net cash income than those of Ko Lao. Table 22 shows that Had Sai Khao families earned a net income of B 1,019 a year, while Ko Lao families had a negative net income, if both cash and non-cash expenses are subtracted from their total cash income. On the average, Had Sai Khao households generated a net cash income of B 16,058 per year (B 1,338 baht per month), while Ko Lao households made only B 6,053 per year.

Borrowing and debt. Indebtedness was very low among Ko Lao village households during 1983. The average debt at the end of the year for households which had borrowed was about B 800. Of the 14 Buddhist households, only two (14 per cent) still had debts, averaging B 100 per household, at the end of 1983, while seven Muslim households (58 per cent) borrowed an average of B 686 per household. About 54 per cent of the borrowed funds were used for household consumption and 45 per cent for costs of production (e.g., buying gasoline, fuel, lubricant, and salt for household fisheries and shrimp-paste making).

Friends or relatives were the only source of loans for households in Ko Lao. No interest was charged; for religious reasons Muslims do not charge or pay interest on loans, and loans to

Buddhist households in this village were also without interest.

Eight out of 26 Had Sai Khao households (31 per cent) reported borrowing money in 1983. The average amount borrowed was about B 2,956. The primary reason for borrowing was to repay debts carried over from the previous year. Households which had borrowed in 1983 repaid 93.4 per cent of the total loans carried over and borrowed in the year (B 6,744). Debts remaining at the end of 1983 averaged B 445.

Funds were borrowed from friends, relatives, and village merchants. The interest rate charged by merchants was generally 20 per cent per year, while friends and relatives charged 15 per cent. The two village merchants who had lent money stated that they had to charge a relatively high interest rate in order to force the borrowers to pay back quickly. In addition they indicated that it is risky to lend to a household in Had Sai Khao because there are no secure assets to take for collateral. Merchants have to take the risk of lending to Had Sai Khao villagers as a part of their business strategy to maintain close links with these households. As a result of taking loans, the borrowers are indirectly forced to sell their fishing catch to the merchants who are the lenders.

Had Sai Khao households spent nearly 90 per cent of the borrowed money on household production and 10 per cent on consumption, in contrast to

TABLE 22.   Average annual income, expenses, and net income per household (in baht)

	Ko Lao			Had Sai Khao
	Buddhists	Muslims	Total	
Average cash income	36,699.9	26,267.7	31,885.1	63,978.9
Average expenses[a]				
for production[b]	19,129.9	13,207.3	16,396.4	19,368.4
	(40.5)	(37.3)	(39.3)	(30.8)
consumption[c]	28,086.0	22,195.0	25,368.0	43,591.0
	(59.5)	(62.7)	(60.7)	(69.2)
total	47,215.9	35,402.3	41,764.4	62,959.7
Net income	−10,516.2	−9,134.1	−9,879.3	1,019.2
Cash expenses	29,203.3	21,899.3	25,832.2	47,920.6
Net cash income	7,496.7	4,368.4	6,052.9	16,058.3

a. Figures in parentheses are percentages of total expenses.
b. Includes estimated cash value of labour and other non-cash inputs to production.
c. Includes estimated cash value of household products consumed at home.

Ko Lao households, which used over half of their borrowings for consumption.

Current economic situation. According to interviews with household heads in both villages, living conditions were difficult. Income was said to be decreasing because of the decline in populations of mature sea animals and the deterioration of the mangrove forests around the islands. They indicated that if this situation continued to get worse they might have to move to places that offered better opportunities for fishing.

Recommendations

The studies of the socio-demographic characteristics, economic situation, and other aspects of life in these two mangrove villages reveal problems that should be alleviated and overcome by the government or concerned institutions in order to assist the mangrove dwellers to improve their living standards before the situation becomes worse. The following recommendations are based on results of the study.

1. An adult-education programme should be established in both Ko Lao and Had Sai Khao villages to assist illiterate persons. There is a relatively high percentage of illiteracy, especially in Ko Lao.

2. Ko Lao households face a problem of population pressure and an economic burden for the families because 50 per cent of the family members are children under age 15. Even though some birth control methods have been used, they have been relatively ineffective. A birth control programme should be implemented in this village to reduce propulation growth in the near future. Because Muslim households were found to have a greater burden of dependency than Buddhist households, the programme should include methods that will not conflict with their religious beliefs.

3. Because capture fishing is the most important occupation of most households in both communities, other fishery operations, for example small-scale, inexpensive programmes like raising fish in cages, should be considered and tested for feasibility. Such projects may give more opportunity for the villagers to work in addition to their sea fishery. This is important because it is expected that the sea fishery around the Ranong estuary will become less profitable and more costly in the near future due to rapid decrease in the supply of marine animals in the area. This implies that the two

mangrove villages should receive more attention from the Department of Fisheries and other related agencies. Demonstrations of aqua-farming techniques at the village level should also be made.

4. Most of the marine animals caught by the fishermen from both villages were sold fresh, since there are no processing techniques other than drying. Shrimp paste was made only by Ko Lao villagers, who use old methods; the quality is low and the process is time-consuming. They receive a low price for it and thus earn a low income. In order to improve both the quality and the quantity of the shrimp paste, some appropriate production techniques should be introduced in Ko Lao and also in Had Sai Khao, where shrimp paste is not now produced. Other fish-processing techniques should be demonstrated in both villages to increase income.

5. The government, through the Royal Forest Department, should provide a definite area of the forest, or a "village forest," to the villagers to allow them to use mangrove wood for firewood and charcoal and timber for house construction. The local forests should be taken care of by the villagers. Villagers and the concessionaires do not co-operate on forest business, and the villagers claimed that destruction of the mangrove forests around both villages was the fault of the concessionaires. They felt that exploitation of mangrove forests by concessionaires was unfair to them. The government should require concessionaires to co-operate with mangrove villagers by hiring them to work in the charcoal kilns, to protect the forest, or to participate in other forest work, so that they can get more income and develop a good relationship with the concessionaires. This will contribute indirectly to the conservation of the mangrove forests around both villages.

6. A supply of fresh drinking water is an urgent need for the two villages. Because the rainy season lasts eight or nine months each year, collection of rain-water would be a cheap source of fresh water. This should be considered as a means to increase the fresh water supply. Most of the households reported that they could store only a little rain-water because they do not have enough water containers. This shortage of fresh water could be eliminated if the villagers were trained to make inexpensive water containers. A programme for this purpose has already been established and is available from the Department of Community Development, and the local government of Ranong Province can request help from this department.

The government may provide two or three water containers to each village for the collection of rain-water for public consumption in case of a sudden severe shortage of fresh water. An investment programme for one or two public dug wells in each mangrove village is also recommended.

7. First-aid health-care services are also needed for Ko Lao and Had Sai Khao, whose residents often face problems of sudden illness. In order to overcome constraints of the local government budget which make it impossible to build and staff a standard health-care station for the small village population, a less expensive health service programme is recommended. A good example of this kind of programme is the village medical bank, and a short-term training programme for one or two villagers to become local village health workers. These workers would be trained in how to give first-aid services to the villagers. Some plants from the mangrove forests might be used as alternatives for mediine, but a scientific assessment of their effectiveness and safety is needed before they can be recommended.

8. A co-operative organization should be set up before the above programmes are implemented in order to increase the co-operation among the villagers, which will be necessary in order to ensure that the programmes succeed.

References

Aksornkoae, S. 1976. *Structure of mangrove forest at Amphoe Khlung, Changwat Chantaburi, Thailand.* Forest Research Bulletin 38. Faculty of Forestry, Kasetsart University, Bangkok.

Aksornkoae, S., P. Iampa, and Kooha. 1982. "A comparison of structural characteristics of mangrove forest near mining area and undisturbed natural mangrove forest in Ranong." In *Proceedings of the National Research Council of Thailand—Japan Society for the Promotion of Science Rattanakosin Bicentennial Joint Seminar on Science and Mangrove Resources, Phuket, August 2–6, 1982,* pp. 149–163. National Research Council of Thailand, Bangkok.·

Aksornkoae, S., S. Priebprom, A. Saraya, J. Kongsangchai, and P. Sangdee. 1984, *Research on the socio-economics of dwellers in mangrove forests, Thailand: Final report submitted to the United Nations University.* Faculty of Forestry, Kasetsart University, Bangkok.

Chaitiemvong, S. 1983. "Shrimp in mangrove and adjacent areas." Paper presented at the Regional Training Course on Introduction to Mangrove Ecosystems, 2–30 Mar. National Research Council of Thailand, Bangkok.

Gedney, R. H., J. M. Kapetsky, and W. W. Kuhnhold. 1982. *Training on assessment of coastal aquaculture potential in Malaysia.* South China Sea Fisheries Development and Coordinating Programme, Manila. (SES/GEN/82/35)

Jintana, U., K. Ogino, A. Komiyama, and H. Moriya. 1983. "Diameter growth measurement by dendrometry." Paper presented at the Mangrove Ecosystem Workshop in Hiroshima, 19–22 Oct. Hiroshima University, Hiroshima, Japan.

Klankamsorn, B., and T. Charuppat. 1982. *Study on changes of mangrove forest areas in Thailand by using Landsat imagery* (in Thai). Forest Management Division, Royal Forest Department, Bangkok.

Kooha, B. 1983. "Litter production and decomposition rates in mangroves adjacent to mining areas and natural mangroves at Changwat Ranong." M.S. thesis. Faculty of Forestry, Kasetsart University, Bangkok.

Martosubroto, P., and N. Naamin. 1977. "Relationship between tidal forests (mangroves) and commercial shrimp production in Indonesia." *Marine Research in Indonesia,* 18:81–85.

Miyawaki, A., S. Sabhasri, S. Aksornkoae, K. Suzuki, S. Okuda, and K. Fujiwara. 1983. "Phytosociological studies on the mangrove vegetation in Thailand." *Bulletin of the Institute of Environmental Science and Technology, Yokohama National University,* 10 (1): 75–111.

National Research Council of Thailand. 1977. *Report of Thailand National Task Force: Mangrove laws and regulations.* Bangkok.

Petchmedyai, J. 1980. "Socio-economic study of mangrove area, Amphoe Khlung, Changwat Chantaburi" (in Thai). M.S. thesis. Environmental Science Programme, Faculty of Forestry, Kasetsart University, Bangkok.

Phuritat, V. 1975. "Relationship between volume and weight of charcoal from *Rhizophora apiculata.*" *Report on research activities,* pp. 38–46. Royal Forest Department, Bangkok.

Royal Forest Department. 1983. "Statistics of management of mangrove forest in Thailand." Unpublished report. Bangkok.

Sarigabutr, D., M. Thummanond, T. Iempairote, and S. Baimoung. 1982. "Climate in southern Thailand." In *Proceedings of the National Research Council of Thailand—Japan Society for the Promotion of Science Rattanakosin Bicentennial Joint Seminar on Science and Mangrove Resources, Phuket, August 2–6, 1982,* pp. 36–46. National Research Council of Thailand, Bangkok.

Sukwong, S., et al. 1976. "Status report on the floristic and forestry aspects of mangrove forests in Thailand." Paper presented at Seminar/Workshop on Mangrove Ecology, 10–15 Jan., Phuket, Thailand.

Vibulsresth, S., C. Ketruangrote, and N. Sriplung. 1976. "Distribution of mangrove forest as revealed by earth resources technology satellite (ERTS-1) imagery." Paper presented at Seminar/Workshop on Mangrove Ecology, 10–15 Jan., Phuket, Thailand.

3. HEALTH AND SANITATION AMONG MANGROVE DWELLERS IN THAILAND

Puckprink Sangdee

Medical Care Available to the Residents of Mangrove Villages

This study of health and sanitation among the residents of mangrove communities in Thailand was carried out in conjunction with the research on Ko Lao and Had Sai Khao villages reported in chapter 2, above. The limited income and low level of education of the average villagers in both Ko Lao and Had Sai Khao affect their living style. Poor hygiene and sanitation create environmental and health problems, impairing the quality of their lives.

Only one household in Had Sai Khao (4 per cent) had a drug cabinet, compared with six Ko Lao households (27 per cent). The drug cabinet in Had Sai Khao was within reach of children and potentially dangerous to them, whereas those in Ko lao were not. None of the drug cabinets in either village were located in such a way as to be exposed to sunlight or other sources of heat.

Despite the existence of drug cabinets, the villagers still do not have enough of the drugs necessary for first aid (Department of Commercial Relations 1982). Drugs commonly found are analgesics, antipyretics, antacids, anti-inflammatory agents, drugs for cough, and antiseptic solutions. Some owners of drug cabinets never get rid of expired drugs, and this practice may be hazardous because the expired drug may become toxic or may have no therapeutic value (Martin and Martin 1978; Milford and Drapkin 1965).

Neither village has a government health centre, nor are there private clinics in the vicinity. A governmental mobile health-care unit makes a brief visit to each of the two villages once a month. The health team provides free primary-health-care services and drugs, including non-prescription medicines and contraceptive pills, and supplies some types of drugs to the principal of the village school, who dispenses them to any villager who needs them. The health team also gives school children basic dental services.

About 44 per cent of the villagers said they were not satisfied with the government health services offered to them. They wanted the doctor to visit more frequently and to investigate and treat of their diseases instead of simply distributing medicine, and they wanted the government to set up a clinic in each village.

One indication of the inadequacy of the health-care services provided by the government is that about two-thirds of the villagers reported they have had to buy drugs from a drugstore to treat their illnesses. About 95 per cent of these said that they bought types and brands recommended by the salesman or the drugstore owner, people who are not trained pharmacists, and the remainder (4–6 per cent) bought drugs as a result of advertisements in various media. Some 23–30 per cent of the villagers have had to go to a private clinic or hospital in town for more serious health problems or for more rapid or frequent service, while the remainder simply neglected mild or tolerable illnesses. Some of them believe that the diseases will go away as easily as they come, and so they never treat their illnesses, and some do not have enough money to pay a doctor. Table 1

TABLE 1. Annual household expenditures for health care

Expenditure (baht)	Households			
	Ko Lao		Had Sai Khao	
	No.	%	No.	%
Under 200	5	19	5	19
200–999	15	58	13	50
1,000–4,999	6	23	7	27
Over 5,000	0	0	1	4
Total	26	100	26	100

This study was supported by the United Nations University.

44

shows the amounts of money spent annually on health care.

Traditional Healing

One Buddhist household in Ko Lao village believes in spiritual healing of illness. For example, the father worshipped his ancestors by offering them food in order to cure his son's illness. They believe that someone in the family may have done something to offend the ancestor spirit.

Some villagers (15–30 per cent) also seek additional help by using traditional medicines, especially medicinal plants which have been identified by Smitinand (1980) and Manoonpiju (1983) (table 2). The effectiveness of these medicines, as reported by the villagers, varies from poor to good, but their pharmacological activity has not been investigated. Villagers usually obtain information about traditional medicines from their relatives and friends. The information may be incomplete or misleading. If they use a traditional medicine incorrectly or misunderstand its therapeutic value, it may be harmful in two ways: first, the treatment may be ineffective or toxic, and, second, it may delay or prevent the patient from seeking more effective forms of treatment.

The four medicinal plants most commonly used are from the mangrove forest. These are two kinds of *ngueak plaa mo* (*Acanthus ebracteatus* and *A. ilicifolius*), two kinds of *samae* (*Avicennia officinalis* and *A. alba*), *thua thale* (*Bruguiera parviflora*), and *seng* (*Triumfetta rhomboidea*). Two other medicinal plants are used by Ko Lao villagers. All these plants can be used in either fresh or dried form. Details of their use are as follows:

— *Acanthus ebracteatus* or *A. ilicifolius* is used for kidney stones. The whole plant is boiled in fresh water, and the patient drinks the solution instead of water, half a glass at a time, until the signs and symptoms disappear.
— *Avicennia officinalis* or *A. alba* is used for thrush in children. The heartwood is rubbed against a coarse stone into fine particles; lime juice is added; and the mixture is stirred vigorously to make a paste, which is spread on the child's tongue twice a day before meals (morning and evening) for three days.
— *Bruguiera parviflora* is used to relieve constipation. The whole plant is boiled in water, and a glassful of the solution is taken twice a day after meals.
— *Triumfetta rhomboidea* is used for fever during menstrual peiods. The whole plant is boiled in a pot of water, and the solution is drunk instead of water, half a glass at a time. Water is added

FIG. 1.   Inadequately treated injuries lead to infections

TABLE 2.   Medicinal plants used by mangrove villagers

Common name	Botanical name	Family	Condition treated	Part used
Ngueak plaa mo	Acanthus ebrac-teatus	Acanthaceae	kidney stones	whole plant
Ngueak plaa mo namngoen	Acanthus ilicifolius	Acanthaceae	kidney stones	whole plant
Waan nam	Acorus calamus	Araceae	kidney stone	aerial parts
Samae khaeo	Avicennia alba	Avicenniaceae	thrush	heartwood
Samae dam	Avicennia officinalis	Avicenniaceae	thrush	heartwood
Chumhet	Cassia alata	Caesalpiniaceae	constipation	whole plant
Ma khaam	Tamarindus indica	Caesalpiniaceae	colds, fever during menstruation	leaves aerial parts
Kameng	Eclipta prostrata	Compositae	first stage of paralysis	
Yaa khaa	Imperata cylindrica	Graminae	fever during menstruation	whole plant
Maiyaraap	Mimosa pudica	Mimosaceae	jaundice	whole plant
Chettamuun phloeng daeng	Plumbago indica	Plumbaginaceae	kidney disease and resultant back pain	aerial parts
Chettamuun phloeng khaao	Plumbago zeylanica	Plumbaginaceae	kidney disease and resultant back pain	aerial parts
Prong	Ceriops tagal	Rhizophoraceae	thrush	heartwood
Thua thale	Bruguiera parviflora	Rhizophoraceae	constipation	whole plant
Krathom	Mitragyna speciosa	Rubiaceae	diarrhoea and stomach ache	leaves
Seng	Triumfetta rhom-boidea	Tiliaceae	fever during menstruation	whole plant
Phlai	Zingiber cassumunar	Zingiberaceae	bruises of internal organs	whole plant

and reboiled until the solution becomes tasteless. This is repeated three times.

Drug abuse is a problem among divers. When they want to dive deeper than the limits their ears can normally tolerate, they take analgesics. They claim that they do not then experience pain in their ears during deep dives. This type of drug abuse is extremely dangerous to their ears, with damage varying from impaired hearing to total deafness.

Types of Illnesses Reported

Villagers were asked to report illnesses they had experienced. Common communicable diseases were reported in both villages with about the same frequency, except that skin diseases were not reported by Ko Lao villagers, whereas tuberculosis and whooping cough were not reported by residents of Had Sai Khao (table 3).

TABLE 3.   Communicable diseases reported in Ko Lao and Had Sai Khao

	Ko Lao		Had Sai Khao	
	No.	%	No.	%
Common cold	26	100	26	100
Influenza	8	31	3	12
Diarrhoea	14	54	4	15
Skin disease	0	0	4	15
Malaria	15	58	10	38
Tuberculosis	1	3	0	0
Whooping cough	3	12	0	0

Numbers and percentages of household heads reporting that the disease occurred in their household.

Malaria is a preventable and usually curable parasitic disease. A full course of drug treatment will usually eliminate the parasites and cure the dis-

FIG. 2. Fungus infection, Had Sai Khao villager

ease. Patients in the two villages treat themselves symptomatically and inadequately. This practice is hazardous not only to their own health but also to others, since the *Plasmodium* organism will ultimately develop resistence to treatment if effective drugs are taken only sporadically. Consequently the inadequately treated patient will harbour resistant strains of malarial parasites that are much more difficult to destroy, and these resistant parasites may eventually spread throughout the community.

Although there is some question about the safety and efficacy of vaccination against whooping cough and tuberculosis (Fulginiti 1982, 134), vaccination programmes against whooping cough (DPT vaccine) and tuberculosis (BCG vaccine) are mandated by the Ministry of Public Health and should be administered to every infant. The occurrence of these two diseases in Ko Lao suggests either that the vaccination programme has failed or that there is a lack of co-operation of the parents, or both. The ineffectiveness of the programme in Ko Lao may be associated with religious faith; these two diseases are not found in Had Sai Khao, where the residents are all Buddhist.

A striking difference between the two villages is fungus disease of the skin, *Tinea versicolor,* which is commonly found in Had Sai Khao, where fresh

water, obtained from a distant well, is usually saved for drinking and cooking (fig. 2). The disease is uncommon in Ko Lao, where fresh water is more plentiful. This disease normally occurs in people with poor skin hygiene and is generally found to be related to the lack of fresh water for washing and bathing.

Village Water Supplies

Each village school has a big reservoir to collect and store rain-water. However, this provides only enough for the children during the school day and for the teachers' families. In spite of the persistent long-term water shortage in both villages, they have not built public reservoirs big enough for everyone in the village. Only three households have one to three large earthenware jars for collection of rain-water for drinking and cooking.

The water supply at Ko Lao, where there are three fresh water wells, is far better than at Had Sai Khao. The first well belongs to the village school and is reserved for the school children and the teachers and their families. The second is privately owned, but the owner allows his neighbours to use it. The last is a public well. The privately owned well has plenty of fresh water all year round, but water from the public and school wells is scarce in the dry season and often has a salty taste. During the summer the owner of the private well limits the time when he allows his neighbours to use it.

Had Sai Khao households have to carry fresh water from a year-round source on nearby Ko Kew island. They transport it by boat in heavy, bulky containers, which is inconvenient and risky.

Birth Control

The birth control programme in Had Sai Khao is quite successful. Both oral and injectable contraceptives are used by housewives in both villages (table 4). The injectable form is more convenient, since only one injection is needed every three months, but a trained person is required to inject the drug. Sterilization has also been used in both villages. In some Moslem households birth control is not used because the family believe it is against their religion.

Sanitary Conditions

The average number of children per family is slightly higher in the Muslim families than in the Buddhist families. The average household income in the villages is quite low compared with the average wage of B 66 per day in Ranong Province (National Economic and Social Development Board 1984, 48), and therefore most parents can provide only inadequate care for their children's well-being. Most children do not have enough clothes

TABLE 4.   Use of modern birth control methods

| | Households[a] | | | |
| | Ko Lao | | Had Sai Khao | |
	No.	%	No.	%
Pills	11	42	10	38
Injections	1	4	1	4
Sterilization	3	12	7	27
Subtotal	15	58	18	69
None	11	42	8	31
Total households	26	100	26	100

a. Households in which one or more women used the method.

to wear, and they play barefooted on the ground. They wear clothing, their school uniform, only when they go to school. These children are at risk for helminthiases because several kinds of worms, including hookworms, *Strongyloides,* and others, can penetrate the human skin (Krupp and Chatton 1982). Toddlers wander around the house and the village without supervision much of the time. They may pick up things from the ground and put them in their mouths, and thus, with poor sanitation throughout the villages, they are likely to swallow worm eggs or larvae. Helminthiases create various health problems, including anaemia, malnutrition,

FIG. 3.   Latrine built over the water, Ko Lao village

and liver diseases, which can impede the children's normal growth and development.

None of the village houses, except the teachers' houses owned by the government, have either indoor or outdoor toilets or lavatories with septic tanks (fig. 3). The lack of facilities for the safe disposal of human wastes, along with the lack of fresh water, seems well correlated with diarrhoea, which is commonly reported in both villages. Poor sanitary conditions may cause epidemic diseases, especially those of the gastro-intestinal system, which can have dangerous consequences.

References

Department of Commercial Relations. 1982. *A manual for public health workers*. Office of the Permanent Secretary of Public Health, Bangkok.

Fulginiti, V. A. 1982. "Immunization." In C. Henry Kempe, Henry K. Silver, and Donough O'Brien, eds., *Current pediatric diagnosis and treatment*. Lange Medical Publications, Los Altos, Calif., USA.

Krupp, M. A., and M. J. Chatton. 1982. *Current medical diagnosis and treatment*. Lange Medical Publications, Los Altos, Calif., USA.

Manoonpiju, K. 1983. *List of Thai plants and references to phytochemical research* (in Thai). 2 vols. Department of Chemistry, Faculty of Science, Mahidol University, Bangkok.

Martin, E. W., and R. D. Martin. 1978. "Distribution and storage factors." In *Hazards of medication*, 2nd ed. J. B. Lippincott Company, Philadelphia, Pa., USA.

Milford, F., and A Drapkin. 1965. "Potassium depletion syndrome secondary to neuropathy caused by outdated tetracycline." *New England Journal of Medicine*, 272: 986.

National Economic and Social Development Board. 1984. *Fact book on labour, employment, salaries and wages*. Wages and Employment Planning Sector, Population and Manpower Planning Division, National Economic and Social Development Board, Bangkok.

Smitinand, Tem. 1980. *Thai plant names*. Forest Herbarium, Royal Forest Department, Bangkok.

4. HUMAN HABITATION AND TRADITIONAL USES OF THE MANGROVE ECOSYSTEM IN PENINSULAR MALAYSIA

H. T. Chan

The mangrove ecosystem in Peninsular Malaysia has played a vital role in the economic and social well-being of the traditional coastal communities. A range of products can be harvested which provide the livelihood of the people living in or near the mangroves. Products include wood for making charcoal, domestic fuel, and construction materials. Locally important industries such as the manufacture of *atap* (nipa "shingles"), nipa cigarette wrappers, and slaked lime are also based on mangrove resources and provide rural employment. Other products harvested from mangrove areas include crustaceans, molluscs, and fin fish.

In recent years, extensive areas of mangroves have been converted for other land uses. Such exploitative conversions often diminish the value of the mangrove ecosystem. The aim of this paper is to document the various types of traditional human habitation in the mangroves and the dependence of their occupational activities on the continued viability of the mangrove ecosystem.

Traditional Human Habitation

Three major types of communities in or associated with the mangrove forests of Peninsular Malaysia can be distinguished: Chinese fishing villages, Malay fishing villages, and other coastal Malay villages. The location of these villages in relation to the Matang mangroves in Perak state (fig. 1) can be taken as representative of their distribution elsewhere in the country.

Chinese Fishing Villages

The Chinese fishing villages (*bagan*) are located within the mangrove forests, often scattered along the banks of mainland or island mangrove estuaries (fig. 2). With the exception of a few well-established mainland villages that are accessible by motorable roads, most of these villages are remotely located and accessible only by river. They vary in size, the larger ones (such as Pulau Ketam

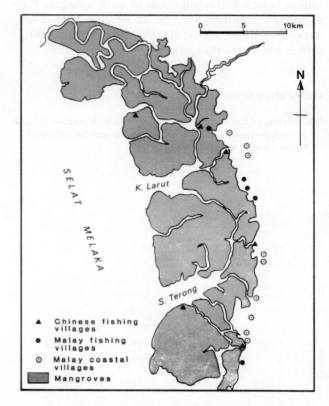

FIG. 1. The location of various types of communities within and along the fringes of the Matang mangroves, Perak, Malaysia

in Selangor) having populations of several thousand inhabitants.

A conspicuous characteristic of these villages is the dense conglomeration of the houses, built on wooden platforms raised on concrete or nibong-palm (*Oncosperma filamentosa*) piles. Adjoining the river-front houses, extensions of the platforms serve as communal jetties, often thickly surrounded with scores of fishing boats. Most of the houses are made from sawn timbers with zinc roofs. Communication within the village is by means of a network of wooden walkways raised on mangrove stilts.

50

FIG. 2.   Bagan Kuala Sepetang, a Chinese fishing village on the banks of the Sepetang estuary in the Matang mangrove. The dense conglomeration of houses built on wooden platforms is a conspicuous characteristic of such villages.

Public electricity and water supplies are lacking; communal generators provide electrical power, while potable water is derived from rain-water collected from roof-tops into large concrete or metal tanks. During prolonged droughts, water has to be purchased from the mainland. Villagers rely on the often turbid, brackish river water for washing and bathing. Charcoal and firewood from the mangroves remain an important source of domestic fuel.

Many of the well established accessible mainland villages, such as Kuala Sepetang and Teluk Kertang in Perak, are almost self-sustaining and possess most of the amenities of towns, auch as shops, community halls, temples, medical clinics, workshops, and schools. Electricity and water supply are readily available, and many of the houses have been rebuilt using concrete.

The majority of the inhabitants of these villages are full-time commercial fishers. They usually fish in waters several kilometres off the mangroves, using a variety of commercial fishing gear. During unfavourable weather, they revert to fishing in the mangrove-fringed rivers and estuaries using traditional fishing gear, or spend their time repairing their boats and nets. To ensure additional income,

some fishers engage in cage culture of fish or crabs. In some villages in Perak, Penang, and Selangor, the culture and harvesting of cockles (*Anadara granosa*) is an important traditional activity.

Malay Fishing Villages

The Malay fishing villages (*kampong nelayan*) are usually found upstream along the banks of mangrove-fringed tidal rivers. The houses are more widely spaced than in the Chinese fishing villages, each having a sizeable compound. They are raised on wooden beams and made from sawn timber. Atap is commonly used as a roofing material; and, to increase living space, the kitchens are often extended to the rear with atap shingles. In villages located some distance from the rivers, raised boardwalks lead to the jetties where the fishing boats are docked.

The occupants of these villages are mainly traditional fishers who confine their fishing activities to mangrove-fringed estuaries. Each fisherman normally operates two or more types of gear that can be used in different localities and weather conditions. To supplement their income, some of them also engage in part-time logging activities in nearby forests.

Coastal Malay Villages

Other Malay villages (*kampong*) are found along coastal roads at the landward fringes of mangroves, with easy access to coastal towns. There is also a cluster of Malay villages, established in the 1930s, remotely located in the middle of Pulau Lumut, a mangrove island in Selangor accessible only by ferry.

The physical structure of these villages is similar to that of the Malay fishing villages except that the houses are surrounded by plantations of cash crops such as coconuts, oil palms, coffee, and cocoa. The villagers are mainly farmers, whose livelihood depends on these agricultural products. Nipa palms (*Nypa fruticans*) commonly grow in the vicinity, and the making of atap shingles is an important occupation.

Traditional Forest Resources

Nipa Industries

Atap Making

The manufacture of atap "shingles" from the leaves of the nipa palm is a traditional part-time cottage industry (Burkill 1935) which remains important among the coastal Malay villagers. Despite the present trend towards replacing timber with concrete houses in the coastal villages, there is still enough demand for atap for thatching roofs and constructing walls and partitions to sustain the industry. The main consumers now are largely charcoal manufacturers and poultry farmers, who still use nipa shingles for roofing their sheds.

The atap maker cuts nipa fronds in the forest with a machete and ties the severed leaflets into bundles, each of which can be made into about 30 shingles. It is common practice to leave the first pair of young fronds on a plant to ensure its recovery from defoliation. The plants are usually harvested on a rotation of four to five months. Occasionally the atap maker employs workers to harvest the fronds, paying them about US$0.80 per bundle of leaflets. A worker can usually collect up to eight bundles a day, thus earning up to US$6.40.

Nipa or coconut (*Cocos nucifera*) leaf stalks, cut

into 1.5 m lengths and spliced into five or six divisions, are used as ribs on which the shingles are fabricated. Groups of two or three leaflets are folded approximately midway over the rib and stitched in place with a strip of peel from the leaf stalks of either nipa or *Donax arundastrum*, a common plant forming dense thickets along the banks of tidal rivers (fig. 3). Freshly collected leaflets are preferred, as they are more pliable and easier to manipulate. The fabrication is usually done by the women of the household, working beneath their stilt houses or under specially constructed atap sheds. It takes about three or four minutes to complete a shingle, and a worker can make up to fifty or sixty a day. Occasionally the workers are employed. In Matang an atap maker pays US$0.10 for ten shingles, or US$0.50–$0.60 a day.

The completed shingles are spread in rows to dry

FIG. 3.   Making an atap "shingle." Two or three nipa-palm leaflets, folded about midway, are stitched into place with a strip of peel from a leaf stalk.

in the sun, usually for about seven to ten days, after which they are tied into stacks of 25 pieces for sale.

The shingles are sold to consumers at about US$7 per hundred. An atap manufacturer can produce up to 2,000 shingles per month, giving a gross monthly income of about US$140. If he employs· labour for leaf collection and fabrication, the net monthly income is reduced to about US$70.

The durability of nipa shingles for roof thatching depends largely on the angle of pitch and degree of overlapping. A high-pitched roof with closely stacked shingles can last up to five years without major repairs.

Cigarette-Wrapper Manufacture

The manufacture of cigarette wrappers from young, unfolded nipa leaf sheaths (Doscas 1972) is a flourishing industry in Peninsular Malaysia, particulary in the districts of lower Perak and northern Selangor. There are essentially two groups of people involved: Malay villagers, who collect, prepare, and dry the leaflets, and Chinese entrepreneurs, who are responsible for bleaching, cutting, and packing and the distribution of the final product.

Young nipa leaf sheaths that have attained a

length of about 1.5–2 m are cut by the local villagers. One worker can usually cut and convey to his depot about 100 sheaths per day. The leaflets are then severed from the stalks with a machete, each cut removing a pair of leaflets just above the point of attachment. About 60 to 80 leaflets, representing about 300 grams of prepared material, can be obtained from a single sheath. The younger leaflets at the tip of the sheath are usually discarded because they are too small and tender. The leaflets are then tied into bundles.

The next process is the removal of cuticle from the leaf blades (fig. 4). This requires special skill and is usually done by women. The worker takes a leaflet and strips one of the blades from the mid-rib with a swift tearing motion. Beginning from the basal edge, she then separates the cuticle sufficiently with her teeth to allow the introduction of a finger, which she quickly forces along the point of attachment, thus completely skinning the blade. The remaining blade with the adhering mid-rib is treated similarly, removing the cuticle and mid-rib together.

The skinned blades are then dried in the sun for a day. The compound used for drying is covered with either defoliated leaf stalks or discarded cuticles, and the blades are placed on top of them in neat rows. During the drying, the material curls slightly, emitting a distinct crackling sound. The

FIG. 4.   Removing the cuticle from nipa leaf blades

dried blades are tied into bundles to be sold to the entrepreneur at about US$53 per 100 kg. The monthly production per household is only about 200 kg.

Most of the dried material is transported to Teluk Intan in Perak, which is the centre for the nipa cigarette-wrapper industry. Here, the bundles or blades are graded and bleached. The bleaching is done with sulphur dioxide produced by burning sulphur granules in specially constructed gas chambers. The process takes two to three hours, and is essential in order to make the blades more pliable and easier to roll into cigarettes. The treated materials are subsequently cut into lengths suitable for smoking, and sold in small bundles or packets.

The consumers of nipa cigarette wrappers are mainly paddy farmers in the northern part of Peninsular Malaysia. The prefer them to ordinary paper-wrapped cigarettes because they resist a certain amount of wetting, a condition that is inevitable during paddy farming operations.

Wood Production

Thinning for Poles

Intermediate felling (thinning) for poles has been carried out in the Matang mangroves since 1930 in forest stands 15 and 20 years of age (Noakes 1952). A one-third thinning of 25-year-old stands, prescribed earlier, was omitted in the present Working Plan (1980–1989), since it led to overcutting and degradation of the residual stand during final felling for either charcoal or firewood production. There are presently about 75 registered pole contractors in the Matang mangroves, and the area to be thinned has been estimated to be about 2,349 ha per year (Haron 1981).

Normally a contractor is allocated two forest areas a year, of about 16 ha each — one for the first thinning (at 15 years) and one for the second thinning (at 20 years). On average, about 3,000–4,000 poles are obtained from the first-thinning area and 1,000–2,000 from the second-thinning area. It has been estimated that about 3,400 trees are left standing per hectare after the first thinning and 1,600 after the second thinning (Haron 1981).

For each area, the contractor employs three or four workers to fell and extract the poles, usually appointing one of them as a foreman to ensure systematic felling and a fair allocation of individual working areas.

The felling is done with axes, starting usually from a selected river bank and working inwards. The process involves selecting a well-formed tree (usually of a *Rhizophora* sp.), and using a stick to describe a circle within which straight trees are to be felled. The stick length is 1.2 m for the first thinning and 1.8 m for the second. The process is then repeated. The felled trees (usually 7.6–12.7 cm dbh) are chopped into suitable lengths (usually 4.9, 5.5, and 6.1 m), using the axe as a measuring tool.

The workers then carry the poles on their shoulders individually to the river bank and stack them to await boat transport to the jetty. Often, walkways are constructed by laying poles two or three at a time end to end, forming a rough track for carrying out the poles. The workers are paid on the basis of the number of poles stacked at the river bank. For trees near the river, they are paid about US$0.40 per pole, and the rate increases progressively to US$0.60 per pole for inland trees. It is common to find inland areas inadequately thinned, particularly where shoulder carrying becomes increasingly tedious, or where the poles have to be transported along creeks in a small boat to the main stacking area. The additional work of loading and unloading often deters workers from working in such areas even though the felling of such inland trees fetches higher wages.

A worker can cut and transport about 30 to 40 poles per day. He normally works for only 15 to 20 days a month since the transport boat is only able to dock at the stacking site during periods of high water. For nearby felling areas, the workers commute daily, while for distant areas, they have to seek accommodation in temporary shelters at the logging site.

When enough poles have been collected, they are transported to the jetty by boats (*tangkang*) with a loading capacity of 200–300 poles. The boatman, who is employed by the contractor, is paid US$0.10 for each pole transported. At the jetty the poles are sold to consumers at a price of US$0.70–$1.50 each, depending on the length and size. With rapid housing development, there is now a good demand for mangrove poles for pilings.

Exploitation for Charcoal

The exploitation of mangrove timber for charcoal production in the Matang mangroves was started in 1930 with the introduction of the charcoal kilns. Since then, it has become the most important form of mangrove utilization (Noakes 1950). There are at present 55 charcoal contractors in Matang, with a total allocation of about 9,000 ha of charcoal coupes.

The contractors are usually given felling areas of about 10–20 ha annually. The felling and extraction of trees are done by a team of usually four or five workers, employed by the contractor. The contractor rarely exercises direct supervision in the working area, though he may occasionally visit it. The work is entrusted to a foreman (kepala), who is responsible for the erection of a barrack and proper division of working areas among the workers.

The barrack is usually a temporary shelter of mangrove poles and atap, with cooking facilities and sleeping accommodation. It is often located beside a creek so that creek water can be used for washing and bathing. Potable water is obtained from rainfall channelled from wooden troughs on the roof into tanks. During the dry season, however, it is necessary for potable water to be brought in by boat.

When the barrack is completed, each worker prepares a stacking platform at a spot on the river bank within his working area where the transport boat (tongkang) can come alongside at high tide. The next step is to construct extraction tracks by laying billets (cut to the right length for charcoal manufacture) parallel to each other at regular intervals, with sawn planks on top to form the track.

The worker then begins the actual felling, using a chain saw and starting with the trees beside the track. When enough trees have been felled (usually ten a day), they are cut into billets 1.6 m long. The bark is removed from the billets, using a wooden mallet made from a Rhizophora prop root. The debarked billets are carried to the stacking platform by wheelbarrow. To assist in lifting and balancing the loaded wheelbarrow, the worker uses a woven rattan strap placed over his neck, with its end loops fixed to the wheelbarrow handles. Normally, a worker takes two days to load a tongkang (with a capacity of 150 pieces), and for this he is paid US$60. He works only for about two weeks a month and can earn about US$420 in that time.

The charcoal kilns are usually constructed in batteries close to a river bank where the tongkang can dock. The batteries are built of sawn timber, mangrove poles, and nipa shingles, and each

FIG. 5. A battery housing charcoal kilns

houses a row of about ten or twelve kilns (fig. 5). The type of kiln presently used is the Siamese beehive kiln, which was introduced to Matang in 1930 by charcoal manufacturers from southern Thailand (Robertson 1940). It is a dome-shaped structure made of bricks, sand, and clay. There are four vents in the vertical wall, and a door for access. It now costs about US$400 to construct a kiln. The average life of a kiln is about 7–10 years if it is constructed on firm ground and regularly used.

On arrival of the tongkang at the battery site, the wood is unloaded and stacked. If the billets have not been debarked in the forest, the charcoal makers employ workers (often women) to debark them at a cost of about US$0.04 each. Billets of less than 10 cm diameter are not debarked. Debarked wood apparently yields a better rate of conversion. The billets are then loaded into the base of the kiln, which is filled by vertical close packing; the dome is left empty. The bottom of each billet is supported on single bricks to ensure complete carbonization. When the kiln is loaded, the entrance door is partially sealed to form the firing aperture during the burn. Normally, small-diameter mangrove billets are used for firing, but because of the increasing difficulty of obtaining an adequate supply, some operators have shifted to using rubberwood or timber offcuts. The firing schedule includes the ''big burn,'' ''small burn,''

and cooling-down periods, and the whole process usually takes 26 to 30 days. The timing of each step is determined by a headman on the basis of the colour and odour of the smoke emitted from the vents.

Each kiln (usually 6.7 m in diameter) requires a charge of about 40 metric tons of green wood per burn. They are normally fired nine times a year. This would mean that each kiln requires 2.8 ha of forest area each year for full operation (Haron 1981). From the 40 tons of green wood about 10 tons of charcoal can be obtained. The present market value of high-grade charcoal is about US$150 per ton.

Exploitation for Firewood

Firewood is harvested in essentially the same way as wood for charcoal, except that the billets are cut into 1.5 m lengths and do not need to be debarked. After the billets are unloaded at the jetty, they are split into two or four sections before being sold. The wood is sold at about US$25 per ton.

Lime Manufacture

Another traditional industry, the manufacture of slaked lime by burning shell fragments (Watson

FIG. 6.   Hauling in mullet caught in the mesh of a gill net

1928), is still practised in the coastal areas of Nibong Tebal in Penang. The shell fragments are collected by villagers from mangrove areas where they are washed up in large quantities by waves and are sold to the lime manufacturers at about US$6 per ton.

At the factory, the shells are mixed with charcoal fragments in a proportion of 2:1. The kiln is a shallow, open structure made of bricks, clay, and cement, ventilated at the bottom by a layer of perforated bricks forming a cavity connected to diesel-powered bellows. Dried coconut or nipa fronds are laid over the perforated floor of the kiln, which is then half filled with the mixture of shells and charcoal, with fronds placed vertically at certain spots to protrude through the mixture. The fronds are lit from above, and, as the fire spreads, the bellows are brought into action. When the mixture begins to glow, more is added until the kiln is filled. The burning process requires about seven or eight hours, and is usually done at night. The resulting lime is slaked (disintegrated by adding water) and put into bags each weighing about 9 kg. The factory workers are paid US$0.18 for every bag of lime produced, and the bags are sold at US$0.60 each.

The slaked lime produced from shells is used mainly by the housing industry as a cheap form of emulsion paint, and by farmers for liming their soils. It is also used as an ingredient for chewing with betel leaves.

Traditional Fishery Practices

Fishing with Nets

Gill Nets

The use of gill nets (*pukat hanyut* or *rantau*) is a popular fishing technique in the mangroves, normally carried out in coastal or estuarine waters (fig. 6). In Matang, the majority of the fishing community in Teluk Kertang is engaged in gill-net fishing.

A gill net consists essentially of a vertical wall of netting which is set so that fish swimming into it become gilled, or entangled by the individual meshes, unless they are small enough to swim through or too large to enter beyond return. The size of fish that can be cauught is naturally determined by the size of the mesh used, which may vary from 5 to 18 cm. The catch consists mainly of mullet and also includes pomfret and penaeid prawns. The operation is carried out daily by a team of two people, usually in the early morning or in the evening. Fishers using gill nets to catch mullet and pomfret can earn an average of US$16 per trip, while those catching prawns can earn about US$20 per trip.

Recently, gill-net operators along the coast of Perak have improvised a three-layered gill net that operates along the same principles as a trammel net. Fish are caught by a central sheet of netting that has an outer armouring of large meshes on both sides. Those fishers who use the three-layered net claim that it is more efficient, yielding a significantly better catch than does the traditional single-layered type.

Casting Nets

Casting nets (*jala*), being simple and versatile, are widely used in mangrove waterways. The net is normally made of cotton or nylon and has a retaining line at the centre. The margin is weighted by a chain of cast-lead rings. The technique of operation is to fling the net out in such a way as to form a shallow bell shape (fig. 7). Fishing with casting nets is mainly on a subsistence level, and the catch usually consists of prawns (especially *Macrobrachium rosenburgii*) and mullet.

Prawn Push Nets

The prawn push net (*rawa*) is a triangular net supported by two light, crossed poles, with a pair of wooden shoes fixed to the ends of the poles to enable the net to be pushed along under the mud (fig. 8). The operator pushes the net for some distance and then shakes the catch into the bag-like end of the net, from which it can be easily transferred into baskets. This technique is used to catch small mysid shrimp, mainly at the seaward edge of the mangroves as the tide rises. The shrimp caught are used primarily for making shrimp paste (*belacan*), a popular condiment among Malaysians.

Fishing with Lines

Baited Lines

Long lines (*rawai umpan*) with baited hooks are used to catch fish in mangrove-fringed rivers. About 300 hooks are attached at regular intervals to a line some 400 m long, whose ends are fixed

FIG. 7. Casting nets are widely used in the mangrove waterways by subsistence fishers

FIG. 8. A triangular push net supported by two light, crossed poles, used to catch small mysid shrimp

to stakes with markers. Prawns and trash fish are commonly used as bait. Fish are caught by this method during periods of low tidal range, and fishers often stay out three or four nights per trip.

Unbaited Lines

Unbaited long lines (rawai tiada umpan) are used mainly for catching scaleless fish such as rays by foul-hooking them. The fishers usually use more than ten lines, each line carrying about 280 short snoods with sharp barbless L-shaped hooks. The lines are set out at dusk, adjoining each other, with weights at the joining points so that they form a low curtain of hooks just above the sea bed. One end is attached to a buoy and the other is retained on the boat. The lines are hauled in at dawn the next day.

Other Forms of Trapping and Collecting

Catching Crabs

Baited traps. Catching crabs with baited traps is an important activity among the traditional Malay fishing people, normally undertaken along the banks of mangrove-fringed rivers and estuaries. Two types of trap are used, the lift net and the collapsible net. The lift-net trap (tangkul ketam) consists of a small square piece of netting stretched out by two diagonally crossed pieces of rattan or split bamboo with sinkers attached, and a wire bait-holder and a rope carrying a float fixed at the junction of the cross-pieces. The collapsible-net trap (bubu ketam) has entrances at both ends, and is stretched by six galvanized-iron hoops. Baits used are chopped pieces of shark, eel, catfish, and ray meat.

The traps can be handled by one person, but the fishers tend to go out in groups. The operation is usually undertaken on a rising tide and ceases when the tide turns, a practice apparently based on the feeding habits of the crabs. A fisher may use as many as 30 or 40 traps, setting them at intervals of several metres and then going back to each in turn and hauling it up quickly to catch any crabs feeding on the bait. The catch consists primarily of Scylla serrata.

Crab hooks. Individuals also catch Scylla serrata in the mangrove forest with crab hooks. The person scouts around the forest to find crab holes, then simply pushes the hook into the hole to catch the crab and pull it out. Usually there is only one crab in each hole.

Gathering Cockles

Gathering cockles (Anadara granosa) from either natural or cultured beds is an important activity in the Chinese fishing communities in Penang, Perak, and Selangor. The greatest development of this industry is in Perak, where about 1,200 ha of the foreshore are under cockle culture (Pathansali and Soong 1958). Harvesting begins when the cockles have attained a marketable size of 24–30 mm. The gear used is a long-handled close-set wire scoop, usually operated by one person, who stands in a boat, extends the scoop as far as his reach allows, and draws it through the mud with a gentle, rocking motion, trapping the cockles, which are then deposited in the boat (fig. 9). One person working for five or six hours can harvest about 10 to 12 bags of cockles, each weighing about 65–70 kg. The collector is usually paid US$1.80 per bag by the fishmonger, who sells it for about US$5.00.

Collecting Other Edible Molluscs

In most mangrove areas, it is not uncommon to find groups of local residents scouting the forest for edible molluscs such as Cerithidia obtusa (belitong), Telescopium telescopium, and T. mauritsii. Most of the molluscs are gathered by hand, though for some burrowing species digging is essential. A collector in Matang can collect about 7–10 kg of belitong per day, which are sold to a middleman at US$0.40 per kilogram.

Along mangrove-fringed coastal mud flats such as those near Kuala Lelangor in Selangor, the collection of bivalves such as Anadara granosa, Orbicularia arbiculata, and Donax spp. during low tide is a common activity. The molluscs are caught by sweeping a wire hand scoop (tanggok tangan) through the mud. To move about the often soft and deep mud flats, villagers have devised an ingenious technique of using light, flat-bottomed troughs, which are manoeuvred by placing one knee on the trough and pushing through the soft mud with the other leg (fig. 10).

Conclusion

It is evident that mangrove ecosystems in Malaysia have been, and are still, used for the extraction of a variety of plant and animal products by traditional methods for the benefit of local people. Continuation of these activities requires that the remaining mangrove areas should be conserved

FIG. 9. A boatload of cockles, scooped up from the mud with a long-handled wire scoop

FIG. 10. Flat-bottomed troughs used for collecting molluscs from the coastal mud flats

and managed in ways that will ensure their productivity. The mangroves of Malaysia are of both ecological and socio-economic value, and the role they play in traditional cultures should be acknowledged by those concerned with the planning and future development of the Malaysian coastline.

References

Burkill, I. H. 1935. *A dictionary of the economic products of the Malay Peninsula.* Crown Agents for the Colonies, London.

Doscas, A. E. C. 1972. "The preparation of cigarette wrappers from nipah palm." *Malayan Agricultural Journal,* 15: 85–86.

Haron Hj. Abu Hassan. 1981. *A working plan for the Matang mangroves, Perak, 1980–1989.* Perak State Forest Department Publication.

Noakes, D. S. P. 1950. "The mangrove charcoal industry in Matang." *Malayan Forester,* 13: 80–83.

———. 1952. *A working plan for the Matang mangrove forests.* Perak State Forest Department Publication.

Pathansali, D., and M. K. Soong. 1958. "Some aspects of cockle (*Andara granosa* L.) culture in Malaya." *Proceedings of the Indo-Pacific Fisheries Council,* 8: 26–31.

Robertson, E. D. 1940. "Charcoal kilns in the Matang mangrove forest." *Malayan Forester,* 9: 178–183.

Watson, J. G. 1928. *Mangrove forests of the Malay Peninsula.* Malayan Forest Records, no. 6.

5. SOCIO-ECONOMIC PROBLEMS OF THE KAMPUNG LAUT COMMUNITY IN CENTRAL JAVA

Ida Bagus Mantra

The Segara Anakan is an estuarine lagoon system in southern Java, close to Cilacap, consisting of many small, mangrove-covered islands in a lagoon environment (fig. 1). There are three villages on the islands — Ujung Gagak, Penikel, and Ujung Alang — together constituting a community called Kampung Laut. This is a fishing area, and most of the population engage in fishing.

At present, the Segara Anakan and its surroundings are in a state of rapid physical and biological change. Siltation of the lagoon and rapid progradation of its shorelines have accelerated as a result of increased sediment yield from inflowing rivers due to rapid soil erosion in their catchments (fig. 2). Together with the reclamation of bordering swamps and some destruction of mangrove forests, this has reduced the viability of the estuarine fishing industry.

Because the islands are surrounded by the lagoon, water transportation is vital. The villages are accessible only by river, tidal channels, or the waters of the Segara Anakan, and dugout canoes or commercial ferry boats are the only means of transportation. Nearly 80 per cent of the households own canoes, which are used for fishing.

Ferry services, operating daily from Cilacap and Kalipucang, call at Mutean, Klaces, and Karanganyar (fig. 3). These ferries, which have been in operation since 1970, have reduced the isolation of the Kampung Laut people. The larger commercial towns in the area are Kalipucang and Pengandaran (15 and 30 km west), Sidorejo (15 km north-west), Kawunganten (10 km north-east), and Cilacap (30 km south-east) (Su Rito Hardoyo 1982).

The Kampung Laut villages cover 7,350 ha, or 73.50 km², of which only 3.6 per cent is used for

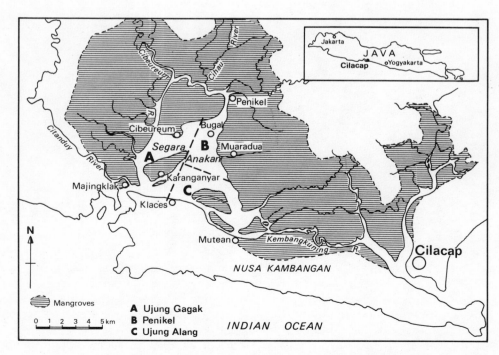

FIG. 1. The Segara Anakan lagoon system in southern Java

62

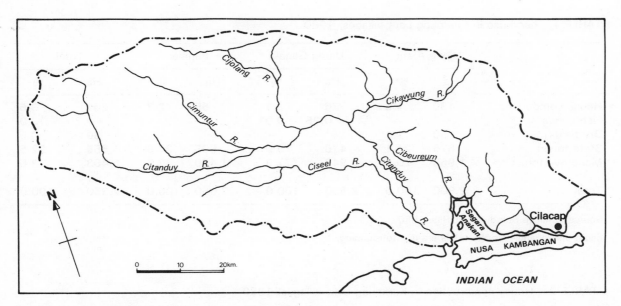

FIG. 2.   The catchment basin of the Segara Anakan. Deforestation and cultivation on steep slopes have caused soil erosion, increasing the sediment yield of the Citanduy river and its tributaries and thereby accelerating the infilling of the lagoon.

FIG. 3.   Ferry boat leaving the village of Klaces, Segara Anakan (Photo: N. Rosengren)

housing compounds; 22.5 per cent consists of state forest, and most of the rest (over 70 per cent) consists of mangroves (table 1).

The villages are compact settlements lining the tidal and river channels or clustering around a central street. Various housing styles and materials are used, with the number of houses built on piles varying between kampungs (table 2). Bugal is the

only kampung where all the dwellings are built on piles. Some 56 per cent of the dwellings are of timber and 37 per cent are of bamboo. The majority have earthen floors. Roofing materials include nipa thatch (32 per cent), tiles (28 per cent), and corrugated iron (40 per cent).

Most wooden dwellings have no separate kitchen, and cooking is done on a small wood-burning

TABLE 1. Land use in Kampung Laut villages, 1980

	Ujung Alang		Ujung Gagak		Penikel		Total	
	Ha	%	Ha	%	Ha	%	Ha	%
House compounds	130	4.6	76	3.1	55	2.7	261	3.6
Fish ponds	—	—	0.25	0.01	—	—	0.25	0.0
Dry fields	30	1.1	—	—	—	—	30	0.0
State forest	675	24.1	476	19.0	505	24.6	1,656	22.5
Mangrove forest	1,965	70.2	1,948	77.9	1,490	72.7	5,403	73.5
Total	2,800	100.0	2,500	100.0	2,050	100.0	7,350	100.0

Source: Cilacap Sub-district Office 1980.

Slight discrepancies in the totals are due to rounding.

TABLE 2. House construction in Kampung Laut villages, 1980

	Ujung Alang		Ujung Gagak		Penikel		Total	
	No.	%	No.	%	No.	%	No.	%
Type								
built on piles	30	20.0	11	11.0	21	42.0	62	20.7
built on the ground	120	80.0	89	89.0	29	59.0	238	79.3
	150	100.0	100	100.0	50	100.0	300	100.0
Floor								
timber or bamboo	30	20.0	11	11.0	21	42.0	62	20.7
earth	101	67.3	66	66.0	27	54.0	194	64.7
concrete	19	12.7	23	23.0	2	4.0	44	14.7
	150	100.0	100	100.0	50	100.0	300	100.0
Outside walls								
bamboo	51	34.0	34	34.0	27	54.0	112	37.3
timber	94	62.7	52	52.0	21	42.0	167	55.7
brick	5	3.3	14	14.0	2	4.0	21	7.0
	150	100.0	100	100.0	50	100.0	300	100.0
Roof								
nipa thatch	49	32.7	26	26.0	22	44.0	97	32.3
tiles	47	31.3	28	28.0	10	20.0	85	28.3
corrugated iron	54	36.0	46	46.0	18	36.0	118	39.3
	150	100.0	100	100.0	50	100.0	300	100.0

Source: Su Rito Hardoyo 1982.

stove or more rarely on liquid- or gas-fuel stoves. Only 1 per cent of the households have bathrooms, and for most people the river and tidal channels provide for bathing and defecation. Consequently, in several kampungs, sanitation is poor and the house and adjacent environment unpleasant (Su Rito Hardoyo 1982).

There is no electricity supply in any of the kampungs, and in the dry season no fresh water supply from streams or tanks. Domestic water is obtained only from springs near Klaces, on Nusa Kambangan island, or occasionally from the Citanduy river. Water must be carried in drums an average of five kilometres to most villages, and up to ten kilometers to some Penikel kampungs. Most houses have small concrete tanks to collect rainwater in the wet season.

Socio-economic Characteristics

The 1980 population census showed Kampung Laut villages as having a *de jure* population of 8,071 residents (3,871 male and 4,200 female) in 1,471 households.

The average number of persons per household was 5.5 — much higher than the 1975 averages in Slemand and Bantul Regencies in the Yogyakarta Special Region, which were 4.4 and 4.5 respectively (Biro Statistik 1974). According to the definition used in the 1971 Population Census of Indonesia (CBS 1976), a household consists of a person or group of persons occupying part or all of a building and generally eating together from one kitchen. The household is the smallest unit in Javanese society, and also forms the basic economic group for production and consumption. In this area a household usually consists of a single nuclear family, but may also include dependent adults and more distant relatives. A newly-wed couple may continue to live with one set of the parents until their own household can be established. Some married couples live permanently with their parents, especially when the latter are too old to work.

The population age structure of the Kampung Laut villages is young. More than 40 per cent of the 1,940 respondents in the recent study by Su Rito Hardoyo (1982) were under 15 years of age, and about 2 per cent were 65 or over (table 3). Age is conventionally taken as an indicator of wage-earning capacity for the purpose of calculating the dependency ratio, the ratio of economically dependent to independent persons in a population. Those between the ages of 15 and 65 are considered independent, while children under 15 and elderly persons 65 and over are considered dependent. The dependency ratio in the Kampung Laut villages was 83.2, which means that every 100 independent persons supported 83.2 dependent persons. This figure is much higher than that reported in the 1980 population census for Indonesia as a whole, which was 75. As the dependency ratio is high, people live at the margin of subsistence. The amount of money earned to supplement food from fishing is very small. This small income is usually used for consumption and other necessities. Mangrove dwellers do not have money for capital investment.

TABLE 3. Age and sex distribution of 1,940 respondents in Kampung Laut villages, 1980

Age (years)	Male	Female	Total	
0–4	141	150	291	
5–9	127	142	269	43.6%
10–14	141	145	286	
15–19	107	122	229	
20–24	70	84	154	
25–29	57	64	121	
30–34	59	66	125	
35–39	66	74	140	
40–44	50	60	110	
45–49	32	28	60	
50–54	22	22	44	
55–59	20	20	20	
60–64	13	23	36	
65–69	12	13	25	1.8%
70+	4	6	10	
Total	963	977	1,940	

Source: Su Rito Hardoyo 1982.

The low income earned by the fishermen in the Kampung Laut villages is reflected in the minimal material wealth of the households. In general a family owns the small household compound on which their dwelling stands. Usually they have a table and chair, some cooking utensils, and a wooden or bamboo bed covered with a mat.

The level of formal education in the villages is very low. Around 60 per cent of the population (4,847 out of 8,071) were illiterate in 1980. Of those who had any formal education, only a few had completed primary school, and even fewer had continued to a higher level of education (table 4). The cost of education is very high and the schools are far away, most of them in Cilacap. The local people see nothing to be gained from a formal education. High school graduates find it very difficult to obtain wage employment. Thus many children do not continue their formal education beyond elementary school, and stay in the village to assist their parents.

As we have noted, fishing is the dominant occupation; it accounts for 88 per cent of the village people with reported occupations (table 5). Farming is a significant occupation only at Klaces and Mutean. Su Rito Hardoyo has pointed out that fishing is arduous in this area, because of the lack of power boats, the difficulty of traversing the

TABLE 4.   Level of education in Kampung Laut villages, 1980

	Ujung Alang		Ujung Gagak		Penikel		Total	
	No.	%	No.	%	No.	%	No.	%
No school/illiterate	2,558	63.9	1,369	52.8	920	62.5	4,847	60.1
Some primary school	1,350	33.7	1,146	44.2	510	34.6	3,006	37.2
Completed primary school	70	1.7	55	2.1	29	2.0	154	1.9
Junior high school	17	0.4	14	0.5	7	0.5	38	0.5
Senior high school	10	0.2	9	0.3	7	0.5	26	0.3
Total	4,005	100.0	2,593	100.0	1,473	100.0	8,071	100.0

Source: Cilacap Sub-district Office 1980.

TABLE 5.   Occupations in Kampung Laut villages, 1980

	Ujung Alang		Ujung Gagak		Penikel		Total	
	No.	%	No.	%	No.	%	No.	%
Fisher	1,939	86.3	1,277	92.3	320	85.6	3,536	88.3
Farmer	238	10.6	—	—	—	—	238	5.9
Trader	30	1.3	25	1.8	24	6.4	79	2.0
Transport worker	19	0.8	37	2.7	14	3.7	70	1.7
Teacher	8	0.4	9	0.7	5	1.3	22	0.5
Other	12	0.5	35	2.6	11	2.9	58	1.4
Total	2,246	100.0	1,383	100.0	374	100.0	4,003	100.0

Source: Cilacap Sub-district Office 1980.

wide mud flats at low tide, and the distances it is necessary to travel to fishing areas. Most fishermen (81 per cent) own their own canoes, which are 5–8 m in length and up to 1 m wide. A variety of fishing gear is used, but most fish are taken in nets or traps (Su Rito Hardoyo 1982).

The best fishing period is between August and December, when the tidal range is more pronounced (spring tides). An average daily catch of 10–15 kg per fishing unit is recorded during this period, compared with 2–5 kg daily during the period January–July (Su Rito Hardoyo 1982).

Most of the catch is sold fresh to a trader in the kampung; there is only a small dried and salt fish industry. The fish trader also acts as a creditor to the fishermen for the purchase of canoes and gear. Repayment is usually made in kind, so that the fishermen may become caught in a debt to the trader and unable to accumulate capital or to sell their catch through another outlet. The distance to the Cilacap market aggravates this condition (Su Rito Hardoyo 1982).

Life-style and Traditions

The Kampung Laut villages are not only places where people live together but also communities whose members believe that they are descended from common ancestors, the Majapahit people. Shared heritage and living together give the villagers a feeling that they are all an inseparable part of the local society, for whose security and welfare they feel responsible.

The principle of mutual self-help, or *gotong royong,* is very strong among the members of the community. They catch fish together or in groups; they co-operate to develop their village; they receive help in the case of a death or other calamity and help others when they are in need of help.

Gotong royong is also practised in connection with socio-religious rituals called *selametan,* whose objective is to maintain harmony between man and nature, man and man, and man and God (Koentjaraningrat 1961). Among these selametan ceremonies is one called *sedekah laut,* held to thank God for his greatness in creating natural resources, and specifically fish resources. People contribute money and foodstuffs for the ceremony; they throw flowers into the sea, and perform shadow plays with leather puppets at night.

Because of the tradition of gotong royong, people are able to survive in these poor Kampung Laut villages.

Problems due to Changes

It has been mentioned that the Segara Anakan and its surroundings are undergoing rapid physical and biological change. The progradation of shorelines and the growth of islands in the Segara Anakan bring about changes in the site properties of the kampung over time.

For those who live in the Segara Anakan and the surrounding area, the siltation of the lagoon is causing several problems. It has been predicted that the lagoon will disappear because of mangrove encroachment and ensuing land reclamation by the early part of the next century. This will result in the loss of the estuarine fishing grounds. As the mangrove forest is reclaimed, the timber resources will disappear, as will the breeding habitats of certain fishes, and a new ecosystem with other biotic components (flora and fauna) will develop (Gembong Tjitrosoepomo 1981). Given these changes and the high rate of population growth, it is important to consider various development programmes for the future.

Conclusion

The Kampung Laut villages in the Segara Anakan region are a complex dwelling area for fishing people. The Segara Anakan lagoon and the surrounding area are undergoing rapid physical and biological change.

The structure of the population is young, so the dependency ratio is high. The small income earned by the fishermen is used mainly for daily consumption and other necessities and none is left for savings and capital. Thus, the people in Kampung Laut villages live at the margin of subsistence. Despite poor economic conditions, they still survive because mutual self-help among the members of the community is strong.

Economic development of the villages has been impeded by the isolation of the settlements from service centres such as Cilacap. Improved water transportation to the larger adjacent towns has been a major factor in increasing the economic opportunities of the Kampung Laut people. Further development of programmes for income and employment opportunities is necessary. Attention must also be given to public health and elementary school facilities.

Several development programmes should be undertaken in the future in response to the gradual siltation and reclamation of the lagoon. Maintenance of the existing environment would require massive dredging and is not economically feasible. It seems inevitable that those who continue to engage in fishing will have to migrate seawards as mangroves encroach on the lagoon, and that eventually their numbers will diminish, their attention concentrating more on marine fishing, while on the reclaimed land their successors will switch increasingly to farming activities.

Acknowledgement

The author is grateful to Dr. Su Rito Hardoyo, lecturer in the Faculty of Geography, Gadjah Mada University, for his kindness in allowing him to use some of his survey data in this paper.

References

Biro Statistik. 1974. *Statistik Daerah Istimewa Yogyakarta, Bagian Tahun 1973.* Biro Statistik Daerah Istimewa Yogyakarta, Yogyakarta.

CBS. 1976. *1971 Population Census, Series G.* Biro Pusat Statistik (Central Bureau of Statistics), Jakarta.

Cilacap Sub-district Office. 1981. *Population Data File.* Cilacap.

Gembong Tjitrosoepoma. 1981. "Pembangunan wilayah pantai Cilacap." Paper presented at a panel discussion on urban and rural planning in Semarang, Central Java.

Ida Bagus Mantra. 1982. "Population and rural settlement in the Segara Anakan region." In E. C. F. Bird, A. Soegiarto, and K. A. Soegiarto, eds., *Workshop on coastal resources management in the Cilacap region,* pp. 86–92. Indonesian Institute of Sciences and the United Nations University, Jakarta.

Koentjaraningrat, R.M. 1961. "Some social anthropological observations on gotong royong practices in two villages of central Java." Cornell University, Ithaca, N.Y., USA.

Su Rito Hardoyo. 1982. "The Kampung Laut of the Segara Anakan: A study of socio-economic problems." In E. C. F. Bird, A. Soegiarto, and K. A. Soegiarto, eds., *Workshop on coastal resources management in the Cilacap region,* pp. 172–182. Indonesian Institute of Sciences and the United Nations University, Jakarta.

6. HUMAN INTERACTIONS WITH AUSTRALIAN MANGROVE ECOSYSTEMS

Eric C. F. Bird

Australian mangrove ecosystems (fig. 1) occupy about 20 per cent of the coastline, and have a total area of about 11,500 km² (Galloway 1982). They extend as far south as Corner Inlet in Victoria (38°55'S), but reach their greatest extent and diversity on the tropical coasts of northern Australia (fig. 2), especially the humid tropical sector of northeast Queensland between Ingham and the

Daintree River. There are at least 30 species of mangroves in the swamps bordering the Daintree estuary. Mangrove ecosystems in southern Australia are generally scrub rather than woodland, the white mangrove (*Avicennia marina* var. *resinifera*) being the commonest, and south of latitude 35° the only, species present. In northern Australia the mangroves include areas of forest, usually with

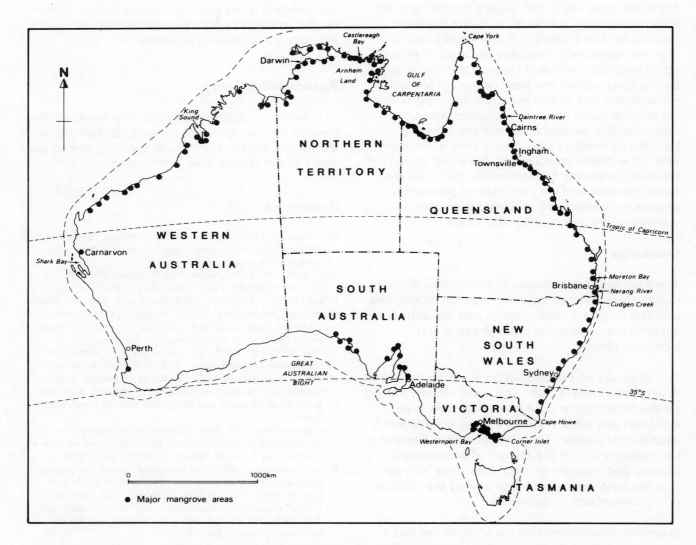

FIG. 1.   Distribution of mangroves around the coast of Australia. Each dot represents the location of a major mangrove area. The dashed line shows the approximate position of the coastline 40,000 years ago.

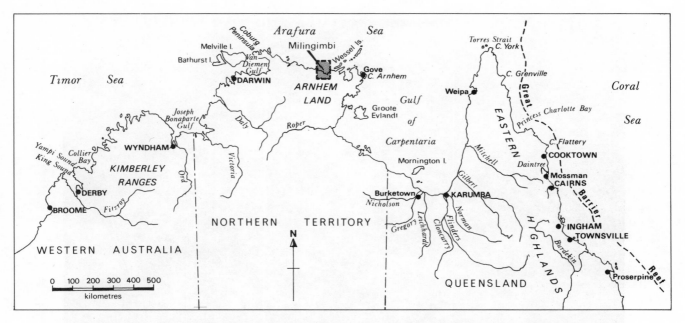

FIG. 2.   Northern Australia (The Milingimbi area, indicated at the top of the map, is shown in detail in figure 4)

scrub communities along their landward and seaward margins, and fringing rivers and tidal creeks (Bird 1972a; Gill 1975).

The greater extent, diversity, and luxuriance of mangrove ecosystems in northern Australia is related to the higher air and water temperatures and to the abundance of muddy sediment in coastal areas, derived from the tropically weathered rock formations of the hinterland and delivered to the coast by rivers. Other factors include the generally large tide ranges, the many sectors protected from strong wave action (e.g. inlets, intricate embayments, and the lee of spits, islands, or coral reefs), and the generally low relief of coastal regions in northern Australia. High salinity is a limiting factor for mangrove growth in the relatively arid parts of north-western Australia, where there are only narrow fringes of mangroves alongside estuarine gulfs such as King Sound. In Shark Bay, on the west coast, sea salinity is too high for mangrove growth. The bulk of the mangrove area is thus on the coasts of the Northern Territory (Wells 1982) and north Queensland (Dowling and McDonald 1982).

Mangroves colonize the upper part of the intertidal zone and spread seaward, usually to about mid-tide level; at low tide they are fronted by exposed tidal flats, muddy or sandy, often with marine plants such as sea-grasses (e.g. *Zostera* spp.)

(fig. 3). While mangroves can grow on rocky, gravelly, or sandy substrates, their best development is on muddy habitats. Sectors of coast receiving muddy sediment prograde as the mangroves spread forward, and there is often a zonation of mangrove species parallel to the coastline (Macnae 1967). It has been shown that mangroves can influence patterns of muddy sedimentation in such a way as to build up a depositional terrace, eventually to high spring tide level (Bird 1972b; Bird, in press). Once this level has been attained, the mangrove vegetation gives place to other ecosystems: in the wetter areas swamp forest, and in the drier areas savanna, salt marsh, or bare hypersaline flats. A feature of many Australian mangrove areas is the presence of cheniers, which are narrow sandy or shelly ridges washed up by storm surges, particularly during tropical cyclones, and left stranded within or behind the mangroves. Hopley (1974) described the emplacement of a chenier in a mangrove area during the storm surge that accompanied Cyclone Althea in the Townsville region in December 1971.

Aborigines and Mangrove Ecosystems

Archaeological research has shown that Aborigines arrived in Australia at least 40,000 years ago, during the Last Glacial phase of the Pleistocene, when the sea level was 100 to 150 metres lower than it is now. Mainland Australia

FIG. 3.   Mangroves spreading onto mud flats at Cairns Bay, north-east Queensland. The pioneer species here is *Avicennia marina,* forming a broad seaward fringe, backed by darker *Rhizophora mucronata,* and then mixed mangrove forest

was then enlarged, and linked by "land bridges" northward to Papua New Guinea and eastern Indonesia, and southward to Tasmania (Jennings 1971; White and O'Connell 1982). Nothing is known of the distribution of mangroves at this low-sea-level stage: the present pattern has developed only within the past 6,000 years as a sequel to the world-wide Holocene marine transgression, which brought the sea up to its present level, submerging Torres Strait and Bass Strait and establishing essentially the modern outlines of Australia (Bird 1984).

Aboriginal tribes who occupied what is now the sea floor around Australia must have retreated as this marine transgression took place. Between 18,000 and 6,000 years ago the sea rose at an average of a metre per century. The rate of land submergence depended on the transverse gradient: off north-western Australia, where the continental shelf is up to 160 kilometres wide, the coastline must have retreated at an average of 13.3 metres per year, whereas off New South Wales, where the continental shelf is only 20 kilometres wide, the average retreat of the coastline was 1.6 metres per year (Bird, in preparation). When the marine transgression came to an end, some Aboriginal tribes remained in coastal areas

(Bowdler 1977), and, as the mangrove ecosystems developed and spread, they made use of them as a place to hunt and collect food items.

The Aboriginal population of Australia is now about 125,000, but only a few thousand Aborigines still live in the traditional manner on the coasts of northern Australia, and even these have modified their life-style in response to European influences. Aborigines have Australian citizenship rights and obligations, and their socio-economic conditions have been transformed by the availability of welfare payments and services and by the purchase and use of retail goods. Particularly in the vicinity of urban centres (such as Darwin, Broome, Cooktown, and Cairns) and mining settlements (such as Gove in north-east Arnhem Land and Weipa on the Cape York Peninsula) Aboriginal culture has been greatly modified; and on the east coast of Queensland the traditional life-style and economy of Aborigines whose predecessors used mangrove areas has been almost forgotten. Even in reservations such as the Yarrabah Mission near Cairns and the Palm Island Aboriginal Settlement off Ingham there is now only incidental use of mangrove areas. South of the Tropic of Capricorn, where mangrove species are fewer and mangrove ecosystems less extensive, there appears to have been little use of mangroves by Aborigines.

It is only in the more remote parts of the north coast of Australia that the traditional use of mangrove areas by Aborigines can still locally be observed. In the Northern Territory, Davis (1984) studied coastal Aboriginal tribes within the region of Arnhem Land, which was legislatively reserved in 1931 for the use of the Aboriginal tribes that had traditionally occupied it. Davis dealt with coastal tribes of the Yolngu Group, which consists of about 30 tribes with a total population of 4,000 to 5,000, living up to 60 kilometres inland.

Some of these tribes inhabit the Crocodile Islands and the shores of Castlereagh Bay, about 450 kilometres east of Darwin (fig. 4). In this remote part of Australia the coastal tribes retain much of their traditional relationship with, and use of, the mangrove-fringed coastline, the main interactions with European Australians having been through the Methodist Milingimbi Mission Station, established in 1921 on Castlereagh Bay, and recently replaced by an independent community with an elected, government-funded Aboriginal Council.

Spring-tide range on this part of the Arnhem Land coastline is about 8 metres, and wide inter-tidal

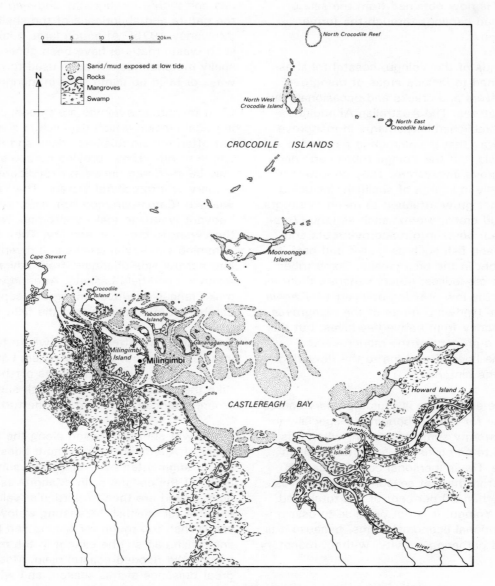

FIG. 4.   Environmental features of the Milingimbi area, Castlereagh Bay, occupied by the Yolngu Aborigines (After a map prepared by Stephen Davis)

areas are exposed and available for foraging as the tide falls. Mangroves fringe sectors of the coastline, especially bordering Hutchinson Strait and alongside the estuaries of the Glyde and Woolen Rivers, and behind these are broad coastal plains, extensively flooded in the wet season, and a hinterland of hilly sandstone country. The Yolngu tribes utilize each of these landscapes, frequenting and using the mangrove areas in the traditional way, drawing upon them for food and other products, but also ranging landward over the coastal plains to the sandstone bush country, and seaward as the tide ebbs. There is no evidence of inland trading of products obtained from the mangroves, and little contact with tourists, although some income is now obtained from the sale of mangrove wood carvings through the former Mission Station.

Within the lands of the Yolngu, coastal tribal estates are defined to include areas of mangrove, bounded by rivers and creeks and occasionally by vegetation features. The Australian Aborigines have never established settlements in mangrove areas in the way that is common in parts of South-East Asia, but the Yolngu tribes certainly use the mangrove ecosystem. They go into the mangroves daily in search of shellfish, including the oysters that grow attached to mangrove roots, as well as mud crabs, worms such as latju (*Teredo* spp.), which bores into mangrove roots and sodden logs, and fish such as mullet and barramundi, caught in the tidal creeks. Some tribes catch and eat crocodiles; others venerate them as sacred. From the low-growing mänyarr (*Avicennia marina*) on the landward fringe of the mangroves they collect honey from native bee-hives, but this is only partly a product of the mangrove ecosystem, as the bees also range to the flowering eucalypts in the hinterland.

The mangrove ecosystem is thus a primary source of food supply for these Aborigines, especially in the wet and stormy summer season, when it is more difficult to catch fish or hunt mammals in the rough sea. Davis (personal communication) estimates that in the wet season the mangroves provide more than 80 per cent of the total food intake of the Yolngu: It is not possible to assess this in conventional economic terms, because it is a hunting and gathering activity, without monetary implications.

In the absence of permanent settlements in mangrove areas, the Aborigines occupy temporary hunting camps for periods of two or three days,

either at the landward margins of the mangroves or on the cheniers, mentioned previously, which provide zones of relatively high and dry terrain. During these forays into the mangroves the older women of the tribe camp on the cheniers and look after the young children, while the young women and older children search for oysters and other shellfish gleaned from the mangrove roots. The shellfish are usually cooked and eaten near the landward margins of the mangroves or on nearby higher ground, and shelly middens of waste material, including charcoal from the fires, have accumulated at these sites. On the Cape York Peninsula some of these middens are mounds up to 20 metres high, accumulated over many centuries and with a stratigraphy showing variations in the nature and abundance of the shellfish utilized (Mulvaney 1975). Although large quantities of such waste material have been generated, the shelly material has not been used to make trackways or build up camp sites in mangrove areas.

Most mangrove swamps are threaded by branching tidal creeks, which diminish in size landward but often remain relatively deep and navigable by canoe or raft. These provide harbours where boats may be sheltered between expeditions out into the estuary or into coastal waters. The Yolngu cut wuduku (*Camptostemon schultzii*), a light and buoyant wood, to make impromptu rafts when they want to cross an estuary. They catch fish by damming small tidal creeks and diverting them into narrow side channels, from which the fish are taken in hand-held baskets. Alternatively, they make fences of woven mangrove saplings that can be set in creeks to trap the fish.

Camp sites are often located where tidal creeks intersect cheniers. On the Arnhem Land coast, men returning from turtle hunts offshore will often take their boats upstream to such sheltered sites, where they can cook and eat the turtles.

In the wet summer season along the northern coast of Australia the maximum tides occur, and these, augmented by river floods, submerge wide areas of the coastal plain. Mangroves adjacent to high ground are then of particular value, being accessible for shellfish harvesting at low tide. The sea is then too rough for fishing and hunting, but some fish can still be caught in the mangroves and on the flooded coastal plain. Mosquitoes are a great nuisance at this season, and when infestations are severe the Aboriginal people coat themselves with a thick layer of mangrove mud, and light smoky fires to drive away the insects.

In the dry season, especially when strong south-easterly trade winds are blowing, the mangroves offer a cool, shady environment in contrast with the wide open coastal plains. Mosquitoes are then less abundant, and Aboriginal people move into the mangroves on hot days to escape heat and dust.

In the dry season, especially when strong south-easterly trade winds are blowing, the mangroves offer a cool, shady environment in contrast with the wide open coastal plains. Mosquitoes are then less abundant, and Aboriginal people move into the mangroves on hot days to escape heat and dust.

Davis (1984) has demonstrated that Aboriginal languages include names for many mangrove species. The most revered is the giyapara (*Rhizophora stylosa*), a stilted mangrove which is seen as embodying the ancestral being who created the coastal area, and which plays a part in ritual and story. Since estuarine shoals are colonized by this species, it is easy to envisage it as a maker of land. Possibly the Aborigines 6,000 years ago, witnessing the cessation of the marine transgression, saw *Rhizophora stylosa* as the mangrove pioneer and derived this legend of stilted mangroves walking in from the sea.

The activities of the Yolngu in the mangroves of Castlereagh Bay, documented by Davis, are similar to those of other coastal tribes. In Western Australia, for example, Kenneally (1982) described the building of fishing platforms by Aborigines in the mangroves. The soft wood of *Camptostemon* is widely used for carvings, some of which are now sold to tourists, and for the making of floats for use with harpoons in hunting marine animals. *Bruguiera parviflora* is used for making spears. Foods from mangroves include the fruits of *Avicennia* and the hypocotyls of *Bruguiera,* while the lemon fragrance of crushed leaves of *Osbornia octodonta* is used for flavouring dugong and turtle meat. Medicinal properties are claimed for a number of mangroves: *Camptostemon* ash is used to treat ringworm and scabies, while *Excoecaria* sap cures skin ailments (Hegerl 1982).

Such traditional use of mangrove ecosystems by Aborigines is thought to have had very little ecological impact, partly because the Aboriginal population has remained small and partly because the techniques of fishing and collecting are still primitive. Conflicts have arisen between Aboriginal and non-Aboriginal commercial fishermen where the

latter have been catching barramundi in coastal waters and thereby depleting the numbers of this fish catchable within mangrove areas by Aborigines. Hostilities, marked by mutual threats and the brandishing of weapons, have given way to discussion and conciliation through legal proceedings. The Aboriginal Land Rights Act (Northern Territory) of 1976 introduced legislation to determine fishing rights and the tenure of inter-tidal and near-shore areas by Aborigines on the Arnhem Land coast (Davis 1984).

Europeans and Mangrove Ecosystems

The various explorers who discovered and charted the Australian coastline — the Dutch in the seventeenth century and the British and French in the eighteenth and nineteenth — described mangroves and included them on their maps, but few of them actually went into mangrove areas. Navigators like Matthew Flinders steered clear of mangrove-fringed coasts, knowing their inshore waters to be shoaly, and it was land-based exploration that led people such as Edmund Kennedy on the Cape York Pensinsula in 1848 and John McKinlay on the coastal plains east of Darwin in 1866 to traverse mangrove swamps. They found them difficult areas, and the settlers who followed them regarded mangroves as hostile environments, habitats for snakes and crocodiles, breeding grounds for biting midges and mosquitoes, waste areas of little or no value until reclaimed as dry land. It is only within the past 20 years that Australian scientists have publicized the ecological importance of mangrove systems and argued for their conservation.

Sundry uses have nevertheless been made of mangrove ecosystems by European settlers and their Australian descendants during the past two centuries. An early use was the harvesting and burning of mangrove wood to produce soda ash, known in Australia as barilla, for soap manufacture. This occurred in the Sydney district in 1810 and also around Moreton Bay near Brisbane, Cairns Bay in north-east Queensland, Port Adelaide in South Australia, and Westernport Bay in Victoria during the early decades of the nineteenth century (Bird 1981). The practice declined only when the industrial production of alkalis by the Le Blanc process, invented in France in 1976 but only slowly disseminated, became established in Australia in the mid-1840s.

In Westernport Bay, Victoria, extensive areas of white mangrove (*Avicennia marina* var. *resinifera,* the only species present here) were cleared and burned to produce barilla in the early 1840s. Records of shipments to nearby Melbourne in 1843/44 have been used to calculate the area of mangroves cleared in that year: 8.4 hectares, or about 7 per cent of the mangrove area in Westernport Bay at that time (Bird 1975). Several of the areas then cleared have failed to regenerate, largely because the clearance of mangroves exposed sandy shores to the rear to erosion by wave action, mobilizing sandy deposits that drift to and fro, preventing the establishment of *Avicennia* seedlings (Bird and Barson 1975).

Mangrove timber has been extracted on a small scale in many areas for jetty construction, boat building, fish traps, mooring poles, and road foundations, and mangrove wood has occasionally been used in carpentry (Hegerl 1982). However, mangrove forest management for sustained yield, based on the systematic cutting of trees and controlled regeneration, has not been practised in Australia in the way that is common in South-East Asia. A minor local industry has been the extraction of tannin from mangrove bark: it is particularly used to make fishing nets more durable. In some areas cattle have browsed mangrove areas or have been supplied with mangrove leaf fodder, usually in times when drought has depleted their usual pasturelands.

Some mangrove swamps have been cleared to make way for salt pans. Areas near Lake Macleod in Western Australia have been converted for salt production, and at Port Adelaide a seaward fringe of mangroves was retained to prevent waves damaging the salt pans.

Estuaries and tidal creeks in mangrove areas have been much used by commercial fishermen and anglers, and mud crabs and oysters are harvested from the mangroves, but fishponds of the kind widely developed in South-East Asia have not yet been contructed in mangrove areas. Salt-water crocodiles, once numerous in the mangrove-fringed estuaries of northern Australia, have been intensively hunted commercially for their skins and are now rarely encountered. For some years they have been declared a protected species, but crocodile farming is now developing as a means of reviving the skin industry while at the same time ensuring conservation of the species (Bustard 1972).

Mangroves have been destroyed in many areas by dumping garbage on them or by deliberate land reclamation. Embanking and filling of mangrove areas has been a prelude to coastal land development, especially in northern New South Wales and southern Queensland, adjacent to resort areas. The cutting of access channels through mangroves to allow boats to reach the land has occurred at many places around Australia — for example at Broome, in the north-west, where such channels were cut to enable pearl luggers and fishing boats to come in and land their catches (fig. 5). Such excavations have had little impact where the dredged material was taken away, but where it has been dumped alongside the channel to form artificial levees, mangroves are killed and replaced by other vegetation (fig. 6). In parts of Westernport Bay the cutting of channels through the mangrove fringe to allow boats to land and collect cattle in the mid-nineteenth century has been followed by a widening of these cut areas, due to the killing of bordering mangroves by drifting sand (Bird and Barson 1975).

In recent years the mechanical excavation of mangrove areas to form marinas for recreational boating has become extensive. In such areas the mangroves have also been modified by the construction of causeways to carry access roads and by the reclamation of land adjacent to the marina for clubhouses and dryland facilities such as boat storage. Similar replacement of mangroves by canal networks, with housing on intervening reclaimed land, has been carried out in several places, for example on the Nerang River estuary in southern Queensland. Here it was found that the canals greatly extended the breeding area for biting midges, which previously bred only in a narrow zone between high neap and high spring tides. The resulting increase in the population of these midges was accompanied by an increasing abundance of mosquitoes, which have a similar breeding habitat, and it has been necessary to introduce widespread spraying of pesticides to combat this nuisance (Hegerl 1982).

At Trinity Inlet, near Cairns, a 726-hectare area of mangroves was embanked in 1972 to be reclaimed and used for sugar-cane cultivation (fig. 7). The damage to mangroves was extensive even outside the embanked area, because the embankment was constructed without regard to the pattern of tidal creek systems, and cut-off loops became stagnant backwaters where mangroves died back. The project has been something of a failure because of an unexpectedly high incidence of flooding from neighbouring steep catchments after heavy rains. It is hoped that any future man-

FIG. 5.   Gap cut through the mangrove fringe to permit construction of a pier —
Broome, in the north of Western Australia, where the mean spring tide range is
8.5 m. Fishing and pearling boats can berth alongside this pier at high tide.

FIG. 6.   The destruction of mangroves by the dumping of dredged sediment
alongside an excavated harbour channel at Yaringa, on the shores of
Westernport Bay, Victoria

FIG. 7.   An embankment built to enclose a mangrove area beside Trinity Inlet,
near Cairns, Queensland, for land reclamation and conversion to sugar-cane farming
(to the left). Note dead mangrove *seaward* of the embankment in areas where the
natural tidal drainage regime was disrupted by this construction.

grove reclamation will proceed only after studies
of the hydrology and ecology of the area have indicated where such reclamation can be successful
and will cause a minimum of disruption to the
mangrove ecosystem.

Dredging in estuaries to improve navigability or aid
flood abatement produces large quantities of mud
and sand, and in many areas this material has
been dumped on adjacent mangroves, killing them.
In Cudgen Creek, New South Wales, dredging of
rutile deposits from the estuary produced large
amounts of overburden that have been dumped in
and around bordering mangroves. Here, as elsewhere, the destruction of mangroves has been followed by a decline in the local fishery.

It is now widely acknowledged that the mangrove
ecosystem is a key part of fisheries ecology,
because mangroves are sheltered, nutrient-rich
areas where many fish species breed and feed.
However, it is difficult to assess this accurately,
because other marine ecosystems, such as salt
marshes and sea-grass beds, also contribute to
the productivity of fisheries, while fish populations
fluctuate independently of habitat factors: a
reduced catch may be due to over-fishing, increased predation, or the onset of a disease. But
mangrove ecosystems, with their high biological

production, rich associated flora and fauna, and
accumulated nutrient-rich substrates, are now considered worth conserving for their scientific and
educational interest as well as their role in maintaining the productivity of fishery resources.

The role of mangroves in trapping and stabilizing
sediment has become obvious in areas where
mangroves have been cleared or have died back
and erosion has ensued (Bird and Barson 1975). A
healthy, spreading mangrove fringe is correlated
with the building up of bordering depositional ter-
races and the maintenance of relatively deep
water channels. When mangroves are cleared or
die away, erosion of their substrate leads to dis-
persal of sediment and shallowing of channels,
thereby diminishing navigability.

Mangroves have been damaged locally by oil spills
and herbicides, as on parts of the shoreline of
Westernport Bay, but the effects of pollution by
urban sewage have been generally to impoverish
or destroy the associated marine organisms rather
than to kill the mangroves. It should be noted that
some cases of mangrove die-back are the outcome of natural changes, such as the in-washing
of sandy deposits by storm surges (fig. 8).

In recent years there has been some planting of

FIG. 8.   Mangrove forest on the south-east coast of Bathurst Island, Northern Territory, killed by in-washed sandy deposits following storm over-wash of a previously protected sand spit (Photo: Stephen Davis)

the declaration of national parks, coastal parks, marine parks, and nature reserves in each of the states of mainland Australia and in the Northern Territory, which include areas of mangroves. However, many mangrove areas remain unprotected, and it is necessary to establish many more reserves that will include mangrove ecosystems with a large number of species, or with features such as zonations that are of scientific interest, as well as those that may contribute to the continued productivity of local fisheries. One difficulty is that mangroves occupy areas submerged at high tide, and it is often simpler to treat them as parts of the marine rather than the terrestrial environment. Yet it is from the adjacent land area that most of the human impacts and influences of mangroves derive.

Acknowledgement

I am grateful to Stephen Davis (Landsearch, Darwin) for information on coastal Aboriginal tribes in Arnhem Land, and for discussion of this paper.

References

Bird, E. C. F. 1972a. "Mangroves on the Australian coast." *Australian Natural History,* 17: 167–171.

——. 1972b. "Mangroves and coastal morphology in Cairns Bay, North Queensland." *Journal of Tropical Geography,* Journal of Tropical Geography, 35: 11–16.

——. 1984. *Coasts.* Blackwell, Oxford.

——. In press. "Mangroves and inter-tidal morphology in Westernport Bay, Victoria, Australia." *Marine Geology.*

——. In preparation. "Man's response to changes in the Australian coastal zone." In K. Ruddle, ed., *Man and Coastlines.*

Bird, E. C. F., and M. M. Barson. 1975. "Shoreline changes in Westernport Bay." *Proceedings, Royal Society of Victoria,* 87: 15–26.

Bird, J. F. 1975. "The barilla industry of Westernport Bay, Victoria." *Victoria Naturalist,* 92: 204–208.

——. 1981. "Barilla production in Australia." In D. G. Carr and S. G. M. Carr, eds., *Plants and man in Australia,* pp. 274–280. Academic Press, London.

Bunt, J. S. 1985. *The mangrove resources of Australia: A country report.* Report prepared for a Unesco/UNDP project. Australian Institute of Marine Science, Townsville.

Bustard, H. R. 1972, "Australian crocodiles." *Australian Natural History,* 17: 149–154.

Bowdler, S. 1977. "The coastal colonisation of Australia." In J. Allen, J. Golson, and R. Jones, eds., *Sunda and Sahul,* pp. 230–237. Academic Press, London.

Davis, S. 1984. "Aboriginal claims to coastal waters in northeastern Arnhem Land, Northern Australia." In K. Ruddle and T. Akimichi, eds., *Maritime institutions in the western Pacific,* pp. 231–251. National Museum of Ethnology, Osaka, Japan.

——. 1985. "Traditional management of the littoral zone among the Yolngu of North Australia." In K. Ruddle and R.

mangroves to restore a vegetation cover that has been depleted or destroyed in the course of mining activities (e.g. near Gove in Arnhem Land), seaport development (e.g. at Gladstone, Queensland), or airport development (e.g. Brisbane, Queensland). In the latter case over 50,000 seedings of *Avicennia marina* and *Aegiceras corniculatum* have been planted adjacent to a runway that extends into the mangroves (Bunt 1985).

Conclusions

Whereas the impact of Aboriginal hunting and gathering appears to have had little effect on mangrove ecosystems in northern Australia, the more widespread and intensive impact of European settlers and their Australian descendants has modified and reduced the mangrove ecosystems during the past two centuries. Dumping of waste materials and deliberate land reclamation have reduced the mangrove area in Australia by about 10 per cent over this period. In recent decades, attempts to conserve mangrove ecosystems have included

E. Johannes, eds., *The traditional knowledge and management of coastal systems in Asia and the Pacific,* pp. 103–124. Unesco Regional Office for Science and Technology for Southeast Asia, Jakarta.

Dowling, R. M., and T. J. McDonald. 1982. "Mangrove communities of Queensland." In B. F. Clough ed., *Mangrove ecosystems in Australia,* pp. 79–93. Australian Institute of Marine Science, Townsville.

Galloway, R. W. 1982. "Distribution and physiographic patterns of Australian mangroves." In B. F. Clough, ed., *Mangrove ecosystems in Australia,* pp. 31–54. Australian Institute of Marine Science, Townsville.

Gill, A. M. 1975. "Australia's mangrove enclaves: A coastal resource." *Proceedings, Ecological Society of Australia,* 8: 129–146.

Hegerl, E. J. 1982. "Mangrove management in Australia." In B. F. Clough, ed., *Mangrove ecosystems in Australia,* pp. 275–288. Australian Institute of Marine Science, Townsville.

Hopley, D. 1974. "Coastal changes produced by tropical cyclone Althea in Queensland." *Australian Geographer,* 12: 445–456.

Irvine, F. R. 1957. "Wild and emergency foods of Australian and Tasmanian Aborigines." *Oceania,* 28: 113–142.

Jennings, J. N. 1971. "Sea level change and land links." In D. J. Mulvaney and J. Golson, eds., *Aboriginal man and environment in Australia,* pp. 1–13. Australian National University Press, Canberra.

Kenneally, K. F. 1982. "Mangroves of Western Australia." In B. F. Clough, ed., *Mangrove ecosystems in Australia,* pp. 79–93. Australian Institute of Marine Science, Townsville.

Macnae, W. 1967. "Zonation within mangroves associated with estuaries in north Queensland." In G. H. Lauff, ed., *Estuaries,* pp. 432–441. American Association for the Advancement of Science, Washington, D. C.

Mulvaney, D. J. 1975. *The prehistory of Australia.* Penguin, Melbourne.

Wells, A. G. 1982. "Mangrove vegetation of northern Australia." In B. F. Clough, ed., *Mangrove ecosystems in Australia,* pp. 57–78. Australian Institute of Marine Science, Townsville.

White, J. P., and J. F. O'Connell. 1982. *A prehistory of Australia, New Guinea and Sahul.* Academic Press, New York.

7. ECOLOGICAL AND SOCIO-ECONOMIC ASPECTS OF ENVIRONMENTAL CHANGES IN TWO MANGROVE-FRINGED LAGOON SYSTEMS IN SOUTHERN SRI LANKA

A. T. Mahinda Silva

Sri Lanka, with a coastline of about 1,700 km (fig. 1), has numerous coastal lagoons, with a total area of about 122,000 ha (Sivakumar 1980). About a third of these lagoons are shallow, with extensively developed tidal flats, mangroves, and salt marshes. The rest are relatively deep, though often fringed with mangrove forest or salt marsh. The mangrove resources of Sri Lanka are largely distributed through these coastal lagoons and on the shores of inlets and sheltered embayments around the coastline. Although people do not live in the mangrove areas, there are many villages and towns close to them. The mangroves have long been used as a source of timber, charcoal, and fuelwood and have provided an environment that sustains fisheries and shellfishing.

A recent research project, sponsored by the United Nations University and structured and carried out by the Marga Institute of Colombo, studied environmental changes, ecological conditions, and sociological aspects of two coastal lagoon ecosystems in southern Sri Lanka (Fernando 1985, 1986). Results of these studies provide a basis for a local assessment of the role and relationships of mangrove ecosystems in the ecology, livelihood, and economic development of people living in this coastal region.

The two lagoon systems are Rekawa and Kalametiya (together with the adjoining Lunama), in the administrative district of Hambantota on the south coast, south of the main coastal highway, east of Tangalla (fig. 2). Rekawa lagoon (fig. 3) is an estuarine system within which, until recently, sea water has been diluted only by relatively small runoff from the land. Kalametiya lagoon (fig. 4) was once similar but has been freshened in recent decades by the inflow of water draining from an irrigated rice area to the north. Lunama is a brackish backwater.

This study was supported by the United Nations University.

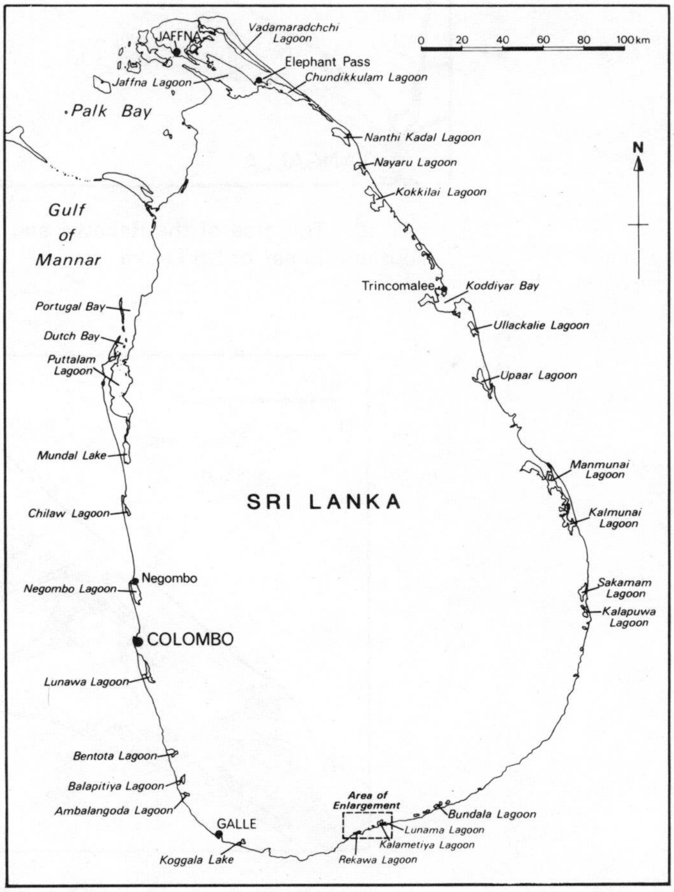

FIG. 1. Coastal lagoons of Sri Lanka. The location of the area shown in figure 2 is indicated at the bottom.

The Setting

Environmental Features

The mean annual temperature in this area is over 26°C, and the average rainfall is often less than 1,200 mm. There are marked seasonal fluctuations in rainfall, and occasional droughts. Soils are chiefly reddish brown earths, but regosols, sandy soils, and silty clay alluvial soils also occur. Under these conditions of climate and soils, the land

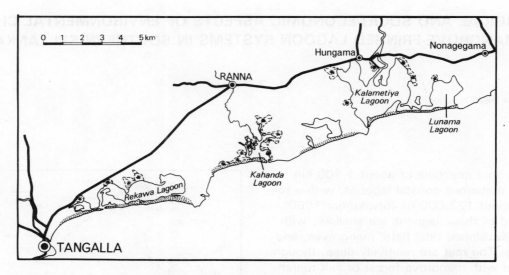

FIG. 2.   The area of the Rekawa and Kalametiya-Lunama lagoon systems on the southern coast of Sri Lanka

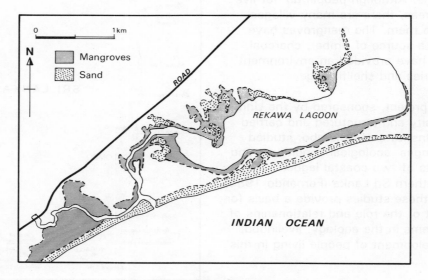

FIG. 3.   Rekawa lagoon

vegetation consists mainly of short, thorny scrub. There are fresh-water wetlands in river valleys, and the lagoons and their fringes have a distinctive swampy vegetation.

The area is accessible only by minor roads and tracks leading south from the coastal highway. Population density is low compared with the better-watered south-west lowlands. Settlements are widely scattered, and there are extensive areas of bare, unused land and thorny scrub. The chief occupations of the people living in the coastal region are agriculture, fishing, and the quarrying of shell grit. *Chena* (a type of slash-and-

burn farming) and irrigated rice cultivation are the most important types of agriculture. With a few local exceptions the standard of living is low.

The infrastructure and service amenities of the area are poor. Flooding is extensive in the lowlands. Some villages do not have sealed roads from the highway, and tracks become submerged and impassable during the rainy season. Provision of clean drinking water is a major problem because in the dry season the local water supply is poor and adversely affected by high salinity. There are a few public water tanks provided by the government, but in general the water supply is

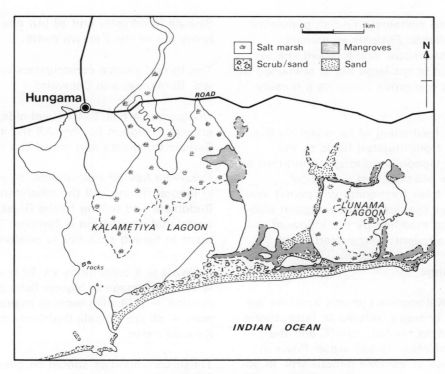

FIG. 4.   The Kalametiya-Lunama lagoon system

inadequate. Another problem is lack of access to schools. Education is hampered by the fact that attendance is irregular, especially when flooding cuts off access.

Mangroves

Studies of the vegetation in and around Rekawa, Kalametiya, and Lunama lagoons together with the associated fauna, especially fisheries, were carried out by S. S. de Silva and M. A. Pemadasa of Ruhuna University (Fernando 1986). Rekawa lagoon is bordered by a mangrove fringe dominated by *Rhizophora mucronata* and *Bruguiera gymnorrhiza*, narrowing toward the entrance from the sea. In Kalametiya lagoon, mangroves form a thin fringe around the eastern and southern shores, passing transitionally to a fresh-water fen vegetation. The mangroves on the eastern shore include *Sonneratia casseolaris*, backed by xerophytic vegetation. In Lunama lagoon, mangroves are confined to the southern shore and the channel linking it with Kalametiya. Species include *Excoecaria agallocha* and *Lumnitzera racemosa*, backed by such xerophytes as *Cissus quadrangularis* and *Cassia auriculata.* In each of the lagoons the mangroves are a narrow, fringing community. In Rekawa and Lunama lagoons they persist by natural regeneration, but it is not yet clear whether they will sur-

vive in the freshened environment of Kalametiya lagoon, where fen and rush swamp may replace them.

Fisheries

There is local subsistence fishing in each of the lagoons, but only Rekawa has a commercial fishery. Both cast nets and kraals, built to trap prawns, are used for fishing. The kraals are temporary structures rebuilt each year, largely from bamboo and mangrove branches. They are used to trap two species of penaeid prawns, *Penaeus indicus*, which is the more abundant and productive, and *Metapenaeus monoceros*, which is present in smaller numbers and only for few months each year. In addition there is an incidental catch of fin fish in the kraals.

The prawn catch varies considerably with seasons and has also shown marked fluctuations from year to year. Variations are related partly to rainfall and runoff regimes and partly to tidal conditions. Biological studies have shown that the two species feed largely on molluscs within the lagoon and its mangrove fringe. Although it is probable that the prawn population is partly sustained by the ecosystem of which the mangroves are a part, this relationship has not been quantified.

Rekawa lagoon also sustains a fin-fish population, notably *Mugil cephalus, Etroplus smatensis, Sarothesodan mossanticus,* and a gobiid species. The fish population is not large but is sustained by the lagoon and its mangrove fringe as a nursery and feeding area.

As a result of the freshening of its water by the increase in runoff from irrigated fields to the north, Kalametiya lagoon is no longer important as a prawn area. It is possible that ecological changes resulting from freshening will permit invasion by fresh-water species. Rekawa lagoon also appears likely to be modified by the inflow of fresh water from a recently excavated canal.

Responses to Change

The Rekawa and Kalametiya-Lunama systems are subject to natural changes, related to fluctuations in such parameters as rainfall, runoff, and the extent of tidal ventilation as sea water flows in through an entrance of variable dimensions. In addition there have been changes due to human intervention, including the recent cutting of a canal to divert water from hinterland paddy fields into Rekawa lagoon, and the earlier, larger-scale diversion of fresh water into Kalametiya lagoon.

The United Nations University – Marga Institute project studied effects of such changes on the socio-economic conditions of six village communities, three around the Kalametiya-Lunama lagoon system and three around Rekawa lagoon (Silva 1986). The significance of the mangrove-fringed system to the activities and livelihood of the local people was investigated in each community.

The three Kalametiya-Lunama communities were Thuduwa, Gurupokuna, and Kalametigoda:

Thuduwa is a community of 74 households, 21 of which engage in local fishing. The total population is 425. The predominant caste in the village is Karawa, a fishing caste, but 11 households belong to other castes. All are Sinhala Buddhists except for one Christian household and two with Muslim men married to Sinhala women.

Gurupokuna has 60 households, of which 44 engage in sea fishing. The total population is 291. The population is entirely Buddhist, and all but one of the households belong to the Karawa caste.

Kalametigoda has 15 households, all of which engage in sea fishing. The total population is 76, all

Sinhala Buddhists, and all but one of the households are of the Karawa caste.

The three Rekawa communities were Godigamuwa, Boraluwa, and Beliwala:

Godigamuwa has 111 households, of which 23 engage in lagoon fishing. All the inhabitants are Sinhala Buddhists and are of the Karawa caste.

Boraluwa has 58 households, of which 19 engage in lagoon fishing. All the inhabitants are Sinhala Buddhists and belong to the Rajaka caste, whose traditional occupation is farming. Their involvement in fishing activities is relatively recent.

Beliwala is a community of 30 households, of which 22 engage in lagoon fishing. With one exception — a Muslim woman married to a Sinhala man — all are Sinhala Buddhists and are of the Karawa caste.

These communities should be considered coastal communities rather than lagoon communities, because their activities are not confined to the lagoons. Coastal communities can be distinguished from agrarian communities and fishing communities in the sense that these terms are normally used and understood. In both agrarian and fishing communities the life of the community has evolved around a major resource that provides its economic base. These coastal communities, on the other hand, depend on several resources in order to spread the economic risks associated with environmental fluctuations and to sustain their livelihood even at marginal subsistence levels. It may not be an exaggeration to say that uncertainty is the only thing these communities are certain of.

In all the communities land ownership is confined to ground that is too elevated for irrigation. The only agriculture suited to the existing climatic conditions in the absence of irrigation is a form of chena known locally as *koratu,* involving intensive cultivation of small cleared areas within the scrub. The extent of scrub jungle has dwindled because of increasing population pressure. By 1960 chena cultivation could be practised in the area only with difficulty. A report on the resources of the area stated, "It is probably over-populated for a dry zone area dependent on chena" (Canada–Ceylon 1960).

The question can then be asked why these coastal communities have not turned to the obviously accessible alternative resource of the sea for their livelihood. In fact participation in sea fishing has been very low in all these communities. The answer to this question leads to a general proposition which appears to be valid, at least in Sri Lanka. It has often been thought that the sea, unlike the land (which is largely held under private or state ownership) is an open resource available to anyone with the capacity and desire to engage in fishing. The reality is otherwise. Fishing communities jealously guard their respective beaches and fishing areas against intruders. The fishing communities are themselves largely closed communities with strong links based on caste and kinship. It is therefore not surprising that very few outsiders have entered the sea-fishing industry (Marga Quarterly Journal 1984).

There is also the further sociological fact that different communities perceive the status of sea fishers in different ways. The geographical location of Thuduwa is such that it is trapped between agricultural communities on the land side, while the lagoon blocks access to the sea. Sea fishing, particularly beach seine fishing and deep-sea fishing, have not been developed by this community as a source of income. Fishing rights in Gurupokuna and Kalametigoda are the sole prerogative of inhabitants of those communities and a few migrant fishermen. In social terms too there is an intra-caste gradation between sea-fishing people and other types of fishing people that has made it difficult for the lagoon fishers to engage in sea fishing. Though of the same fishing caste, Thuduwa fishers concede that the sea fishers are "maha Marakkalas" (big Marakkalas) while they are just Marakkalas. The emphasis is on "maha," because the sea fishers deal in bigger varieties of fish, while other fishers deal in smaller fish for which there is a poor demand.

In Rekawa too, the non-participation of lagoon fishers in sea fishing is due to social constraints. Most respondents in Beliwala and Boraluwa stated they could work only as crew members, subordinate to the boat-owning sea fishermen, although they possessed the necessary skills to engage in sea fishing. The Godigamuwa kraal fishers attributed their non-participation in sea fishing to status differences in the occupations. The kraal fishers regarded themselves as superior to sea fishers on the grounds that, despite the higher incomes associated with sea fishing, the activity is concentrated within a few months of the year and for the rest of the year the sea fishers have little to systain them. They regarded the kraal as an as-

set yielding a more stable income throughout the year.

At present the pattern of employment is such that there is a great emphasis on fishing as a source of income. For Thudawa and the three Rekawa communities the most important source is lagoon fishery. All respondents in Rekawa engaged in fishing in the lagoon during the prawn season, which lasts for four moths, beginning in December and reaching its peak in January. In Kalametiya there was no clearly delineated prawn season, and the prawns caught were of a commercially less valuable variety.

The price paid to the producer of commercially salable prawns is in the range of 67.50–79.50 Sri Lankan rupees per kilogram, while the poorer variety is in less demand and is priced around Rs 27 per kilogram. The highest average monthly incomes are recorded in the prawn season. In Boraluwa and Beliwala most of the annual income of fishing households is realized during the prawn season. Lagoon fish account for most of the income of Godigamuwa households. Income from fishing in inland reservoirs accounts for most of the income of Thuduwa households. In all four communities the number of household members engaged in employment other than fishing is low.

Daily wages for agricultural labour are Rs. 15–25 for males and Rs 15 for females (much less than the price of a single kilogram of prawns). During land preparation and harvesting seasons these wages may rise to Rs 30 and Rs 25 for males and females respectively. Labour is grouped into two broad categories: agriculture-related work and fishing. Agriculture-related work in Rekawa means work in irrigated fields in the region outside these villages. Some people commute to these fields daily during the periods when labour is required. They also may obtain wages for labour in coconut estates further in the interior. This type of work is not available regularly throughout the year. In Boraluwa, the figure for those engaged in agricultural work is higher (83 per cent). The predominant work is koratu, the intensive cultivation of small plots, often of less than half a hectare. Since land for cultivation is scarce in the village, those who engage in this type of cultivation rent from landowners in the hinterland.

Even though at present lagoon fishery accounts for a considerable part of the income of the Rekawa communities, in the long term their dependence on lagoon fishing is likely to be only a temporary phase, as other resources become accessible. For Thuduwa the phase of great dependence on lagoon fishing is over, because Kalametiya lagoon has undergone several changes beyond the control of the local inhabitants and its fish resources have been severely depleted. At Rekawa the lagoon fishery resources may also undergo changes due to factors beyond the control of the people, forcing the Rekawa communities to increase their dependence on other sources of income such as agricultural labour.

Fluctuations in the degree of dependence are not confined to the lagoon resources, as can be seen in the Kalametiya communities. Until the 1950s the lagoon played only a marginal role in economic activities of the community. At that time the three communities of Thuduwa, Kalametigoda, and Gurupokuna had semi-subsistance economies based on shifting cultivation. Agriculture rather than fishing was the main source of livelihood. The vast tracts of scrub that surrounded the villages provided ample land for cultivation. Lagoon products had a very limited market, mostly confined to the agricultural communities of the coastal region.

Cultivation could not continue as the main source of livelihood, however, because a growing population began to claim increasingly large tracts of farmland for settlement. Rapid population growth in these areas was associated with welfare measures implemented by successive governments, the more important among which were the anti-malaria campaign, better health-care services, improved transport, and government-subsidized food. As mentioned, a report on the resources of the Walawe-Ganga basin in 1960 concluded that the area was probably over-populated for a dry zone dependent on chena.

The net outcome of these processes was that peasants, most of whose income had until then been derived from farming, were forced to increase their use of alternative resources, which previously had played only a marginal role. One such alternative was sea fishery, the exploitation of which had up to then been in the hands of outsiders with access to capital and technology in the form of craft, gear, and skills.

Lack of capital was a definite constraint that hampered the local inhabitants from taking on large-scale sea fishing. At this point a coincidence of government policies geared to develop and modernize the sea-fishing sector acted in the favour of

Gurupokuna and Kalametigoda. A principal feature of the government policies was that "in selecting the beneficiaries, government chose workers who were not owners of the traditional boats themselves. The decision had major consequences for the social hierarchy and the structure of power as it had evolved through the traditional fishing based as it was on the traditional technology" (de Silva 1977).

Although the processes outlined above worked to the advantage of the Kalametigoda and Gurupokuna communities, they had the opposite effect on the Thuduwa community. Thuduwa's disadvantageous geographical location made access to beach-seine sites and harbour facilities difficult. The gradual acquisition of use rights brought about the development of marine fishing in Gurupokuna and Kalametigoda. The restrictive regulations on the clearing of scrub for chena cultivation reduced Thuduwa to dependence on two highly seasonal sources of income: lagoon fishing and farm labour. The lagoon fishery proved to be short-lived, being the victim of unintended consequences of a national agricultural policy. The construction of the Walawe Right Bank Scheme and its system of reservoirs and irrigation channels resulted in a large volume of residual fresh water from the newly opened irrigated lands draining into Kalametiya lagoon. Although a canal has been built at the marine entrance to allow water to escape to the sea, the continuous flow of this water requires frequent clearing of the sand bar to prevent inundation of paddy fields located in the upper reaches of the lagoon.

As a result of these changes, the communities of Gurupokuna and Kalametigoda have merged and enjoyed a moderate prosperity through development of marine fisheries, whereas Thuduwa remains less economically developed.

At present 65 per cent of the fishers in Thuduwa derive their income from inland fisheries, but access to this source is becoming increasingly difficult because local communities around the inland reservoirs have organized themselves into "fresh-water fisheries extension societies," one of whose major aims is to limit the fishing rights in these reservoirs to their membership.

Today the communities in Rekawa are confronted with a common issue, the threat of the loss of the lagoon fishery, which is their most important source of income. Construction of a canal into the lagoon to drain an adjacent paddy tract of about

37 hectares, known as Thangalu Welyaya, has already caused changes in the lagoon ecology. These may explain the very low prawn catches during the past season. The communities that depend on the lagoon have complained bitterly about the new canal, but, given the costs of canal construction, its closure is an extremely remote posssibility. The fishing community is not politically strong enough to overrule the agricultural community.

Data for Kalametiya and Rekawa lagoons show that open access to fishery resources cannot be maintained while the resources are economically viable. Rekawa lagoon has never been an "open-access" resource. The Godigamuwa community in Rekawa has monopolized the best sites for kraaling for several generations. Prior to the 1970s, when the demand for lagoon fish and prawns was low, access to the lagoon outside the kraaling area was open to local inhabitants who used cast nets for fishing. The cast net posed no threat to the kraal fishermen, whose technology was comparatively superior. With the increase in population and the non-availability of other sources of income, net fishing has increased steadily.

In the latter half of the 1970s, with the development of the tourist industry and the export market for prawns, demand resulted in prices for prawns rising even higher than for bigger sea fish. With the new marketing conditions, the traditional monopoly of the kraal fishermen has been challenged in an indirect way through use of a more efficient mode of fishing, the drift net. Prior to the late 1970s drift nets were used to some extent, but only to catch fin fish, and the minimum legal mesh size of the nets was 4.2 cm. The increased demand has attracted a large number of fishers to the profitable prawn industry. Almost all of them use illegal nets with a mesh size of about 2.5 cm. Disputes between krall fishers and drift-net fishers have resulted in tighter government regulations. The number of kraals, their specifications, the number of drift nets and their mesh sizes, the numbers of fishers entitled to operate kraals and drift nets are all regulated, but enforcement of the regulations has been ineffective and has not reduced the scramble for a greater share of the fishery.

In Kalametiya the reverse process has occurred. Though the lagoon fishery in Kalametiya lagoon has been virtually destroyed, a fisheries extension society has been established to limit exploitation of what is left of the resource to a group of Thuduwa fishermen. Extension societies have been established in inland fishing communities also, and fishing rights on the reservoirs are now the prerogative of the members of these societies. The absence of any tenurial arrangements in Kalametiya lagoon has made it difficult to enforce the limitation of fishing rights to members of the society.

New market conditions for prawns and lobsters during the latter half of the 1970s revealed gaps in the traditional tenurial system, on which the inhabitants of these coastal communities have been quick to capitalize. One such new opening is lobster fishing. The traditional sea tenure system revolved around beach-seine, off-shore, and deep-sea fishing. Lobster fishing is treated separately because it does not require a particular site. All that is necessary is gear and some means of navigation in the shallow sea. Most fishers use either dugouts or inflated motor-vehicle tyre tubes for the purpose.

Conclusions

One important implication of this study for policy-planners is that undesirable consequences can follow from a lack of understanding of the ecological and socio-economic unit for planning integrated development. An overall development plan embracing a whole region and taking into account diverse interrelated factors is needed to avoid uneven or skewed development, achieved at the expense of certain sections of the population.

This study also highlights the need for development planners to recognize interacting ecological and socio-economic factors involved in the implementation of a development project. The absence of such perceptive planning is seen in the way in which the lagoon resources of the coastal communities of Rekawa have been adversely affected by the irrigation outflow canal for the development of the Thangalu Welyaya tract, which was built despite the fact that the probability of a harmful ecological change was obvious from a glance at Kalametiya lagoon.

Throughout this study it has been evident that the mangrove fringes of these lagoons have not been of major interest or importance to the coastal people, even though they may play a significant role in the lagoon ecosystems. Mangroves do not form extensive landscapes or habitats for human occupation in Sri Lanka. They are but one aspect of a diverse environment, exploited in many ways by the economically marginal communities living in this coastal region.

References

Canada–Ceylon. 1960. *Report on reconnaissance survey of the resources of the Walawa Ganga Basin.* Canada–Ceylon Colombo Plan Project, pp. 107–109. Government Press, Colombo.

de Silva, M. W. A. 1977. "Structural change in a coastal fishing community in southern Sri Lanka." *Marga Quarterly Journal,* 4: 47–86.

Fernando, S., ed. 1985. "Selected lagoon systems in the Hambantota District of southern Sri Lanka." Unpublished report.

Marga Institute, Colombo.

——. 1986. "Use and management of coastal lagoons in southern Sri Lanka," *Marga Quarterly Journal.* In press.

Marga Quarterly Journal. 1984. "Fishery." Special issue. Vol. 7, nos. 2 and 3.

Panayotou, T. 1984. "A resource sector with a difference: Competitive marketing in Sri Lankan fisheries." *Marga Quarterly Journal,* 7: 15–28.

Silva, A. 1986. "Socio-economic aspects of coastal lagoon communities in southern Sri Lanka." *Marga Quarterly Journal.* In press.

Sivakumar, J. 1980. "Evolution of coastal sand and vegetation resources of the north-eastern coastal belt of Sri Lanka." In J. I. Furtado, ed., *Tropical ecology and development,* pp. 109–117. International Society of Tropical Ecology, Kuala Lumpur.

8. THE DISTRIBUTION AND SOCIO-ECONOMIC ASPECTS OF MANGROVE FORESTS IN TANZANIA

J. R. Mainoya, S. Mesaki, and F. F. Banyikwa

Mangrove vegetation is characteristic of sheltered coastlines in the tropics. Mangrove communities are extensive in protected shallow bays and estuaries, around lagoons, and on the leeward side of peninsulas and islands. In Tanzania mangrove forests occur on the sheltered shores of deltas, alongside river estuaries, and in creeks where there is an abundance of fine-grained sediment (silt and clay) in the upper part of the inter-tidal zone.

The establishment of mangrove vegetation is governed to some extent by the degree of exposure to strong winds. The largest continuous mangrove areas are to be found on the coasts of Tanga district in the north, the delta of the Rufiji River in Kilwa and Lindi districts, and in Mtwara, where the Ruvuma River forms an estuary close to the Mozambique border (fig. 1). Thus, the mangrove forests stretch along coastal districts from Tanga to Mtwara and cover an area of 79,937 ha. Mangroves are also well represented on the coasts of the main islands, Zanzibar, Pemba, and Mafia. On Pemba mangroves cover an area of 12,146 ha, while on Zanzibar there are 6,073 ha under mangroves.

Flora of the Mangroves of Tanzania

Studies of the mangrove vegetation of the East African coast include those of Graham (1931) along the Kenya coast, and Walter and Steiner (1936) along the coastal district of Tanga in Tanzania. MacNae (1968) reviewed the literature on mangroves of eastern Africa and included a description of the various plant communities and their zonation. Graham (1931) described the species present in mangrove swamps in Kenya, and commented on their ecology and economic utilization.

McCusker (1971), working on the mangrove vegetation of the Kunduchi area near Dar es Salaam, recorded six principal mangrove species: *Sonneratia alba* Sm. (Sonneratiaceae), *Rhizophora mucronata* Lam. (Rhizophoraceae), *Ceriops tagal* (Perr.) C. B. Robinson (Rhizophoraceae), *Bruguiera gymnorrhiza* (L.) Lam. (Rhizophoraceae), *Avicennia marina* (Forsk.) Vierh. (Avicenniaceae), *Xylocarpus*

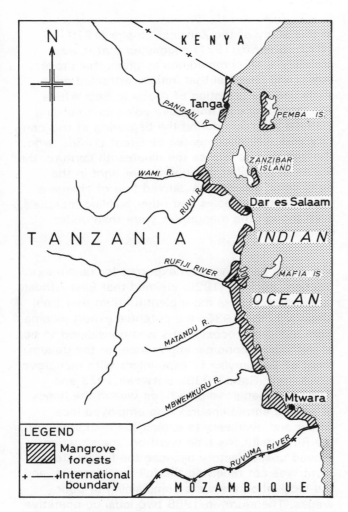

FIG. 1.   Mangrove forests along the coast of Tanzania

S. Mesaki and F. F. Banyikwa are in the Department of Zoology and Marine Biology of the University of Dar es Salaam, Dar es Salaam, Tanzania.

This study was supported by the United Nations University.

granatum Koen. (Meliaceae). In addition, *Heritiera littoralis* Dryand (Sterculiaceae), an "associated species," has been reported from estuarine mangrove swamps in East Africa by MacNae (1968). Other "associated species" frequently encountered in Tanzania are *Lumnitizera racemosa* Willd. (Combretaceae), *Arthrocnemum indicum* Moq., *Salicornia pachystachya* (Bunge ex Ungernsternb), *Sesuvium portulacastrum* Linn., and *Suaeda monoica* (Forsk. ex J. F. Gmel).

The composition of mangrove vegetation has been influenced by a number of factors. For example MacNae (1963) noted that the presence of sand appears to restrict the growth of certain species, notably *Bruguiera, Rhizophora,* and *Ceriops,* and that *Bruguiera gymnorrhiza* predominates under conditions of fresh-water influence. MacNae and Kalk (1962) also noted that *Ceriops* and *Bruguiera* seedlings develop only in the shade of other trees. Though *Rhizophora* can germinate and grow anywhere in the upper inter-tidal zone, it will grow to maturity only in waterlogged areas. Moreover, *Sonneratia alba* is most commonly found in loose muddy sand, and *Xylocarpus granatum* is virtually confined to sandy soils with a low humus content (McCusker 1971).

It has been observed that the usual pioneer species in sandy habitats is *Avicennia marina,* and in muddy areas *Sonneratia alba.* Because of its high tolerance of dry conditions, *A. marina* has a wider distribution than *S. alba,* which is mostly found in silty waterlogged sites. *Heritiera* is much less tolerant of strongly brackish water conditions and is therefore found mostly around the upper limit of sea water in river estuaries.

Walter and Steiner (1936) described the zonation of mangroves in the Tanga region and observed that *Bruguiera gymnorrhiza* does not form a distinct zone but occurs interspersed with *Rhizophora* and *Ceriops.* They concluded that the occurrence of *B. gymnorrhiza* is not directly related to distance from the outer and inner limits of the mangrove swamp but to depth and salinity of the water and the texture of the substrate. The composition of the mangrove community is largely a function of tolerance to salinity of the water and waterlogging of the substrate (Chapman 1970). However, such factors as clearing, changes in amount and regime of rainfall, and changes in the pattern of sand bars at the mouth of a creek which limit the rise and fall of the tide in the mangroves can all lead to changes in the species composition of the vegetation. For instance McCusker

(1971) noted that the clearing of mangroves in Kunduchi Creek resulted in an increase of *Ceriops tagal* at the expense of other species in the regenerating plant community.

Socio-economic Aspects of Mangroves

Overview of Commercial Exploitation

Mangrove poles and bark have been exported from the eastern coast of Africa for a very long time. Trade between the Persian Gulf and East African ports dates back to a time when the Arabs secured a footing on the coast of East Africa. For centuries, dhows from the Gulf visited Lamu, Mombasa, Zanzibar, and the Rufiji delta during the north-east monsoon (*kaskazi*), bringing with them dates, carpets, salt, and earthenware and taking back mangrove poles (*boriti*) and firewood on their return journey during the south-west monsoon (*kusi*).

In this study of the Zanzibar Empire (1770–1873), Sherriff (1971) suggested that it was probably such commodities as grain, mangrove poles, and charcoal that initially attracted and sustained the attention of Arabs to East Africa. The importance of mangrove poles as a building material in Arabia during the beginning of this century was well documented by Grant (1938), who showed that as late as the nineteenth century, the Sultan of Zanzibar retained a user right in the Rufiji delta, whence he derived free of charge a large number of poles and other building materials, even though the mangroves were then under German control.

Mangrove bark has been exported for tannin extraction. Buckley (1929) claimed that East African tannin extract was more plentiful than that from Malaya. In the 1930s the potential export income from mangrove ecosystems was considered to be of sufficient economic importance for the government of Tanganyika to experiment with mangrove silviculture (Grant 1938). Between 1923 and 1958 the mangrove-pole trade was in the hands of private entrepreneurs, who employed local labour and overseers to exploit the mangroves of the Rufiji delta. As time went on, the situation proved unsatisfactory because the Forest Department was not receiving the full royalties due, and the local labourers were complaining of poor wages. Therefore, in 1958 two local co-operative societies, the Kishoka and the Rufiji Delta Co-operative, were formed to exploit the entire Rufiji delta mangrove area jointly. These co-operatives contracted out most of the cutting of mangroves but did not have much control over the marketing of mangrove poles, which mainly benefited the overseers. In an effort to overcome this problem, the co-operatives were reorganized in 1960 to consist only of *bona fide* local mangrove cutters, and three new co-operative societies were formed: the Amini Co-operative Society, the Mbwera Co-operative Society, and the Kiasi Co-operative Society.

This trade (see tables 1 and 2) flourished up to 1960, when a decline set in, probably due to the economic changes that followed the discovery and large-scale exploitation of oil in the Arab Gulf countries.

In 1965 the pole trade went into the hands of the Coast Region Co-operative Union, which handled the trade on behalf of the three co-operative societies, but in 1976 all co-operatives were banned by the government and the mangrove trade was handed over to a para-statal body known as the Tanzania Timber Marketing Company, which still handles all the trade in mangrove poles with foreign countries. In 1979 alone, the company earned 4 million Tanzanian shillings in foreign exchange after selling 30,000 scores of mangrove poles (1 score = 20 poles), two-thirds of which were destined for Iran (Harnevik 1980).

Commercially Exploited Mangrove Species

There are seven species of mangrove of commercial importance to the local people:

1. *Rhizophora mucronata,* the most useful and abundant species of mangrove, is found as almost pure stands, or with scattered trees of *Bruguiera* and *Ceriops,* in muddy tidal areas. Under local conditions, it normally attains a height of about 8 m. It can be easily distinguished from other tree species by its aerial, bowed, stilt roots, many of which arise from quite high on the trunk and branches, and by its viviparous seeds. The bark is more fibrous, thinner, and rougher than that of *Ceriops,* and the tannin content is reported to be in the range of 24.8 to 43 per cent. The wood when seasoned becomes very hard and durable. The poles are used locally as telephone poles, and the wood is used as fuel.

2. *Ceriops tagal* is a shrub or small tree of variable size, growing to a height of 4 m. It is found

TABLE 1.   Number and estimated value of mangrove poles used locally and exported from Tanzania, 1950–1967

	Used locally[a]		Exported	
	Number	Value (TSh)	Number	Value (TSh)
1950	590,189	633,001	565,500	126,200
1951	1,197,000	1,304,730	422,000	104,600
1952	561,600	622,144	346,000	518,820
1953	496,800	541,512	284,300	449,200
1954	383,400	117,906	221,480	386,080
1955	657,600	716,784	155,500	169,495
1956	446,900	487,321	186,000	367,780
1957	366,000	368,950	169,000	159,240
1958	617,760	673,358	166,000	180,940
1959	490,000	534,100	658,600	717,874
1960	—	—	754,000	773,060
1961	—	—	211,000	268,960
1962	—	—	323,751	324,840
1963	—	—	214,514	254,400
1964	—	—	85,937	96,980
1965	—	—	80,475	132,140
1966	—	—	152,310	180,300
1967	—	—	72,563	70,080

Source: Forest Department Annual Reports.
a. Not recorded after 1959.

interspersed with *Rhizophora* and *Bruguiera,* and may be recognized by the angular character of its long viviparous radicle. Economically this species used to be the most important source of non-fibrous mangrove bark (average 40 per cent tannin content). The poles and wood are used locally for building houses and for firewood. Grain-sifting buckets (*nyungo*) are also made from the split stems of this species. The sap from the seeds is used for water-proofing fishing lines.

3. *Bruguiera gymnorrhiza* is less common than *Rhizophora mucronata* but is found interspersed with it. Under suitable conditions it will grow to a height of 20 m. The wood is used as fuel for lime burning and sometimes for building poles.

4. *Sonneratia alba,* a small, common tree, is one of the first invaders of mud flats subject to tidal inundation by salt water but does not extend to the higher levels. The wood is used as fuel for lime burning and occasionally for building poles, and its vertical shoots are used as floats for fishing nets. It is not an economic source of bark tannin.

5. *Heritiera littoralis* is a tall tree with slender,

TABLE 2. Mangrove bark exported from Tanzania, 1923–1967

	Tons		Tons	Value (TSh)[a]		Tons	Value (TSh)
1923	7.3	1938	3,032	—	1953	354	131,200
1924	247	1939	3,596	—	1954	557	211,460
1925	3,097	1940	4,759	—	1955	545	143,220
1926	6,087	1941	6,383	—	1956	300	73,480
1927	8,330	1942	7,291	—	1957	451	106,580
1928	8,909	1943	3,205	—	1958	488	186,300
1929	5,711	1944	5,100	—	1959	1,652	431,000
1930	5,396	1945	4,927	—	1960	1,586	501,060
1931	4,052	1946	5,490	—	1961	577	158,540
1932	660	1947	30	—	1962	1,335	292,440
1933	3,115	1948	1,689	—	1963	919	200,740
1934	2,151	1949	1,446	—	1964	755	268,020
1935	3,715	1950	860	244,800	1965	1,763	523,860
1936	2,973	1951	828	343,480	1966	368	100,940
1937	4,966	1952	967	384,940	1967	880	175,600

Source: Forest Department Annual Reports.

a. Value not recorded before 1950.

TABLE 3. Local uses of primary species of mangrove found in Tanzania

Botanical name	Vernacular names	Uses
Rhizophora mucronata	mkoko, mkaka, msinzi	building poles, bark for tannin, fuelwood, grain-sifting baskets
Ceriops tagal	mkandaa, mwekundu, mkandaa wa pwani	building poles, bark for tannin, fuelwood, fencing
Bruguiera gymnorrhiza	mshinzi, msinzi	building poles, telephone poles
Avicennia marina	mchu	trunks for canoes, carts, masts, bedsteads, chairs, handles
Sonneratia alba	mlilana, mpira, mpera, mkopa, mkoko mpya	fuelwood, construction timber, shoots for fishing-net floats, dhow masts
Xylocarpus granatum	mkonafi, mtifi	dhows, furniture
Heritiera littoralis	msikundazi, mkokoshi, mkukushu	trunks for dhow masts, construction timber, furniture, boats

Source: Temu 1975.

pointed germinating rootlets that attain a length of at least 30 cm and become anchored in the mud. It is used locally as a source of fuel and building materials: poles and rafters. The bark is of little economic value for the extraction of tannin.

6. *Avicennia marina* is a common spreading tree, usually found on the higher levels of swamps. It is willow-like in general appearance and has a light yellowish green foliage. Vertically pointed pneumatophores arise in great abundance from the long, spreading, horizontal roots. The bark is smooth and greenish yellow when young, and variegated green and reddish in the older trees. The trunks are used for making small dug-out canoes, and the tree is generally used for building

carts, for dhow and canoe fittings and masts, for furniture such as bedsteads and chairs, and for fittings such as handles. It is also used extensively as fuel for lime burning. The tree is of no use as a source of tannin.

7. *Xylocarpus granatum* is a small tree, growing up to 6 m in height, that is scattered through the mangroves, especially in the higher parts. The large, round woody fruits, 15–20 cm in diameter, are of medicinal value. The wood is very hard and said to be like teak. It is used for carts, dhows, furniture, and construction. Although the bark contains up to 32 per cent tannin, the tree is considered a poor producer of tannin.

Exploitation of Mangrove Products by the Local Community

In addition to the well-known export trade in poles, mangrove forest products are put to many uses locally (table 3).

Poles (boriti). Straight, round, barked poles between 8 and 16 cm in basal diameter and between 5 and 7 m long, cut mainly from *Rhizophora, Ceriops,* and *Bruguiera* species, are widely used in constructing houses. The mangrove wood, being tough and termite-resistant, is preferred for this purpose by coastal people. It is estimated that over 600,000 poles are used annually to build and repair ordinary Swahili houses in Dar es Salaam (fig. 2).

Mangrove bark. Tannin extracted from mangrove bark is used locally by leather-processing companies since the importation of tanning and dyeing agents has been restricted. In recent decades the question of establishing a tannin-extraction factory to process mangrove bark locally has been discussed, but, although a mangrove-bark factory based on the Rufiji mangroves would be practicable and viable, the government has not yet made a decision.

Charcoal. Tanga, Dar es Salaam, Kilwa, and Zanzibar provide major markets for charcoal. Mangrove wood can be burned to produce excellent charcoal of high calorific value, but difficulties of transport to the centres of charcoal consumption have reduced the profitability and attractiveness of large-scale charcoal production from mangrove forests.

Fuelwood. Nearly all rural people as well as the ur-

FIG. 2. Building poles of mangrove wood on sale in Dar es Salaam

ban poor in Tanzania depend almost exclusively on fuelwood, charcoal, and kerosene as energy sources for cooking and lighting (Nkonoki 1981). People in coastal villages use mangrove wood for firewood in their homes (fig. 3). Mangrove wood can also be used as fuel in coastal-village industries for the production of burnt bricks and tiles and in lime burning where alternative sources of fuelwood are not available.

Timber. According to Grant (1938), the best mangrove timber is *Heritiera* (*msikundazi*), which is the only species commonly found in large dimensions. But the use of large mangrove trees for timber has not been explored in Tanzania because timber from inland tree species is readily available.

Other uses. The coastal people of Tanzania use mangrove products for many other purposes as well. These include boat hulls, masts, and oars, fences for pig pens, and fish traps. In addition, they use *Avicennia* foliage as fodder for goats and cattle and *Rhizophora* leaf extract as a medicine for hernias.

The Importance of Mangrove Areas to Fisheries

Mangrove swamps support a resident fauna dominated by molluscs and crustaceans (Macintosh 1983). The fauna of Tanzanian mangroves has not been studied comprehensively, but it is known that the mangrove waterways and the areas submerged at high tide support important fish populations. The mangroves, especially around the Rufiji delta, may be seen as serving two fishery-related roles: as a habitat and as a nursery ground for many species of shellfish and fin fish that can be exploited commercially.

"Shrimp" of the mangrove and associated saline waters (*Acetes* sp. and *Macrobrachium* spp.) are not extensively exploited commercially. However, many women and children are involved in catching

FIG. 3.   Mangrove wood cut for firewood on sale in Dar es Salaam

Acetes, which may be important to the rural economy.

Estimates of the total prawn potential of Tanzania, based on a survey by Japanese team (Marcotrade 1978) showed that up 2,000 tons of prawns could be harvested each year. This figure is probably an underestimate because the mangrove habitats north of Bagamoyo (Ruvu estuary), south of the Rufiji delta, and on Zanzibar and Pemba were not taken into consideration.

Sankarankutty (1974) identified six penaeid species at Kunduchi Creek: *Penaeus indicus* (Milne-Edwards), *P. monodon* (Fabricius), *P. semisulcatus* (De Haan), *P. japonicus* (Bate), *P. latisulcatus (Kishinouye),* and *Metapenaeus monoceros* (Fabricius). The commercially important species are *P. indicus, P. monodon, P. semisulcatus,* and *M. monoceros.*

In his study of the crabs of Tanzania, Hartnoll (1975) mentioned the crabs of the mangroves around Dar es Salaam. The common mangrove crab *Scylla serrata* is little caught in Tanzania, the estimated total annual catch being less than 100 kg (Sichone, personal communication). This small catch is from the mangrove stands close to the urban consumer centres of Dar es Salaam, Tanga, and Mtwara. There is, however, a great potential for crab fishery in the large mangrove stands of the Rufiji and Ruvu estuaries. For example, the Rufiji delta alone is thought to have a fishery potential of about 20 tons of live crabs per month (Marcotrade 1978).

The common mangrove molluscs of *Terebralia* spp. and *Cerithedea* spp. are not utilized at present in Tanzania, although MacNae (1974) reported that they form an important food source in some countries.

The Need for Defined Mangrove Conservation Strategies in Tanzania

Human activities are by far the most important factor influencing mangroves negatively in Tanzania today. One such activity is the clearing of land in and around mangrove forests to create salt evaporation pans (fig. 4). Tanzania produced 60,000 tons of salt from coastal salt pans in 1981.

Although saltworks have been in production for

FIG. 4.   Salt pans at Kunduchi, north of Dar es Salaam

several decades around Kunduchi, north of Dar es Salaam, significant reduction of the mangrove vegetation in this area was not noticed until 1970 (McCusker 1971). By the end of 1981, however, there had been a large expansion of the saltworks, which reduced the once large Kunduchi Creek mangrove forest to a mere vestige (fig. 5). If this trend continues unchecked, the increasing demand for table salt in Tanzania is bound to produce a corresponding decline in mangrove habitats (Mwaiseje and Mainoya, in press).

Conservation of the mangrove habitats and resources they contain is very important. The island of Mafia and the Rufiji delta mangrove forest have the potential to be a very rich marine park area. The Rufiji delta supports dugongs and crocodiles, both of which are threatened species, and some islands around Mafia support the green turtle (Robertson 1968). In the Jozani forest on Zanzibar, the red colobus monkey, *Colobus badius,* has been observed to feed on the buds and young leaves of the mangrove *Rhizophora mucronata* (Mturi, unpublished observations 1983).

Tanzania has an excellent reputation for wildlife conservation on land; it would be very desirable if the government took a greater interest in mangrove forest conservation. It has been shown that unplanned cutting of mangrove forests often leads to coastal erosion and sediment mobilization, and this adversely affects marine life dependent on the nutrients associated with the mangrove ecosystem. The destruction of the mangroves may reduce not only coastal fishery resources but also catches from off-shore fishing, far from the mangrove forest itself (Macintosh 1983).

Mangroves protect tropical shores from erosion by tides and currents, and Macintosh (1983) recommended that a mangrove strip at least 100 m wide should be left as a buffer zone on the more exposed shores. Although mangrove areas could also become important for aquaculture, it is fortunate that mangrove areas are not utilized for this purpose in Tanzania.

If silviculture is to be practised in mangrove areas, greater consideration should be given to those mangrove species that have been over-exploited for poles, firewood, and charcoal — *Rhizophora mucronata, Ceriops tagal,* and *Bruguiera gymnorrhiza* (Grant 1938). Furthermore, because regeneration of mangroves is a slow process, thinning of mangrove stands could be carried out where necessary to reduce tree density and promote growth of the remaining trees.

The rapid growth of the human population in Tanzania is likely to increase the negative human impact on mangroves. The important role of

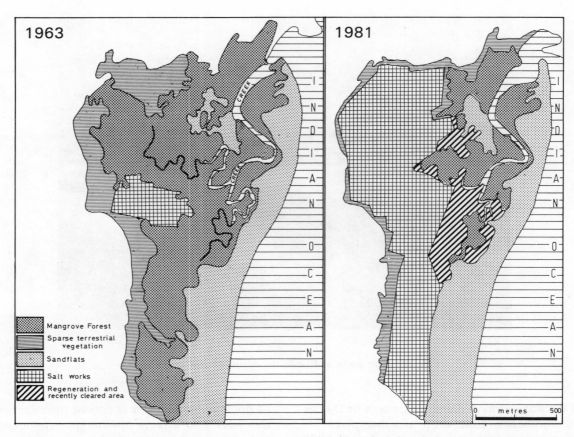

FIG. 5.   Vegetation and land use in the vicinity of Kunduchi Creek, 1963 (left) and 1981 (right)

mangroves in the preservation of coastal and off-shore fish and shellfish stocks and the shore-protecting properties of mangrove vegetation make the preservation and sound management of mangrove-clad habitats an urgent matter for developing coastal states like Tanzania.

Conclusions and Recommendations

1. Mangrove communities in Tanzania have not been studied in sufficient detail to plan their rational exploitation. There is an urgent need for further studies of these ecosystems.

2. Few integrated multidisciplinary studies of mangrove forest use have been made. Most studies have been of a reconnaissance nature, generally providing scanty information that cannot be used in land-use management plans. It is recommended that a more ambitious study of the mangrove ecosystems of the Tanzanian coast be carried out, with a view to providing a data base from which the government and other land-use planners can design their land-use programmes.

3. Because the trade in mangrove poles is very lucrative both for the internal market and for export, it is recommended that a plan for monitoring mangrove pole harvesting and utilization should be worked out. A strategy for growing the economically more important mangrove species could be investigated.

4. A body drawing members from the Ministry of Lands, Natural Resources, and Tourism, the Ministry of Housing and Urban Development, and the State Mining Corporation should be formed to monitor the rational utilization of mangrove areas as a natural resource.

5. The Tanzanian government should ensure that adequate records are kept with regard to the utilization and marketing of mangrove products.

Acknowledgements

The authors would like to thank the United Nations University for its financial support and keen interest in this project. The authors wish also to thank Dr. A. K. Semesi, Mrs. V. Kainamula, and Mr. L. B. Mwasumbi for their assistance and Mrs. Z. Juma for typing the manuscript.

References

Buckley, T. A. 1929. "Mangrove bark as a tanning material." *Malay Forest Record,* no. 7.

Chapman, V. J. 1970. "Mangrove phytosociology." *Tropical Ecology,* 11: 1–19.

——. 1976. *Mangrove vegetation.* J. Cramer, Vaduz (Lubrecht and Cramer, Monticello, N.Y., USA).

Graham, R. M. 1931. "Notes on the mangrove swamps of Kenya." *Journal of the East Africa and Uganda Natural History Society,* 36 (1929): 157–164.

Grant, D. K. S. 1938. "Mangrove woods of Tanganyika Territory: Their silviculture and dependent industries." *Tanganyika Notes and Records,* 5: 5–16.

Harnevik, K. J. 1980. *Non-agricultural activities, Rufiji District: Survey, analysis and recommendations.* BRALUP Service Paper 80/2. University of Dar es Salaam.

Hartnoll, R. G. 1975. "The Grapsidae and Ocypodidae (Decapoda: Brachyura) at Tanzania." *Journal of Zoology* (London), 177: 305–328.

McCusker, A. 1971. "Ecological studies of an area of mangrove vegetation in Tanzania." Ph.D. thesis, University of Dar es Salaam.

Macintosh, D. J. 1983. "Wetlands of the world: Riches lie in the tropical swamps." *Geographical Magazine,* 60 (4): 184–188.

MacNae, W. 1963. "Mangrove swamps in South Africa." *Journal of Ecology,* 51: 1–25.

——. 1968. "A general account of the fauna and flora of mangrove swamps and forest in the Indo–West Pacific region." *Advances in Marine Biology,* 63: 73–270.

——. 1974. *Mangrove forests and fisheries.* FAO/UNDP, Rome. (IOFC/DEVE/74/34)

MacNae, W., and M. Kalk. 1962. "The ecology of the mangrove swamps at Inhaca Island, Mozambique." *Journal of Ecology,* 50: 19–34.

Marcotrade. 1978. "Report on a biological study of crustaceans in the Kilwa and Rufiji areas." Marcotrade–Tanzania fishing venture report (unpublished).

Mwaiseje, B., and J. R. Mainoya. In press. "Mangrove habitats: Problems of conservation in Tanzania." In *Proceedings of the International Society for Tropical Ecology* (Silver Jubilee Symposium, Bhopal, India, 1981).

Nkonoki, S. R. 1981. *The poor man's energy crisis.* A report of the Tanzania Rural Energy Consumption Survey, 1981.

Robertson, I. B. J. 1968. "Marine parks and reserves." Report for the Ministry of Agriculture and Co-operatives, Dar es Salaam.

Sankarankutty, C. 1974. "Coastal and estuarine aquaculture: A case for introducing prawn culture in Tanzania." In Msangi and J. J. Griffin, eds., *International Conference on Marine Resources Development in Eastern Africa,* pp. 26–28. University of Dar es Salaam.

Sheriff, A. M. H. 1971. "The rise of a commercial empire: An aspect of the economic history of Zanzibar (1770–1879)." Ph.D. thesis, University of London.

Temu, E. M. 1975. "Brief notes on the mangroves of Tanzania." In file no. FD/37/17, Ministry of Natural Resources and Tourism, Dar es Salaam.

Walter, H., and M. Steiner. 1936. "Die Ökologie der ostafrikanischen Mangroven." *Zeitschrift für Botanik,* 30: 65–191.

9. SOCIO-ECONOMIC ASPECTS OF MANGROVE VEGETATION IN JAPAN

Akira Miyawaki

Mangroves are typically tropical vegetation, but they extend into subtropical environments in southernmost Japan, growing well on Okinawa and the Amami Islands and reaching their northern Japanese limit on southern Kyushu. They have been discussed in papers by Miyawaki (1980, 1981, 1984) and Miyawaki, Okuda, et. al. (1983).

Mangrove Vegetation in Japan

Mangroves grow on silt and mud flats in bays, estuaries, and lagoons, and along the shores of estuaries. Sites suitable for mangrove growth are very restricted in south-western Japan, and are mostly on the Ryukyu Islands (figs. 1 and 2), where mangroves cover only a limited area. They grow up to 5–8 m in height on Iriomote, Ishigaki, Miyako, and Okinawa islands (Miyawaki, Suzuki, et al. 1983; Miyawaki, Okuda, et al. 1983). Japanese mangrove forests have up to 12 species, including

Rhizophora stylosa, Kandelia candel, Bruguiera gymnorrhiza, Lumnitzera racemosa, Avicennia marina, Sonneratia alba, Excoecaria agallocha, Acrostichum aureum, Nypa fruticans, and *Heriteria littoralis.* At the margins one can also find *Clerodendron inerme, Derris trifoliata,* and *Dalbergia candenatensis.*

Mangroves in Japan are at the northern limit of their distribution in the Indo-Pacific region. They occur as fragmental outliers, typically with only one or a few species, and form scrub or low forests with only one or two vegetation layers. Six mangrove communities have been recognized, dominated respectively by *Avicennia marina, Sonneratia alba, Kandelia candel, Rhizophora stylosa, Bruguiera gymnorrhiza,* and *Lumnitzera racemosa* (Miyawaki 1980, 1981).

On Iriomote the mangroves are poorly developed

FIG. 1.   Mangroves (*Bruguiera gymnorrhiza*) bordering the Miyara River, Ishigaki Island

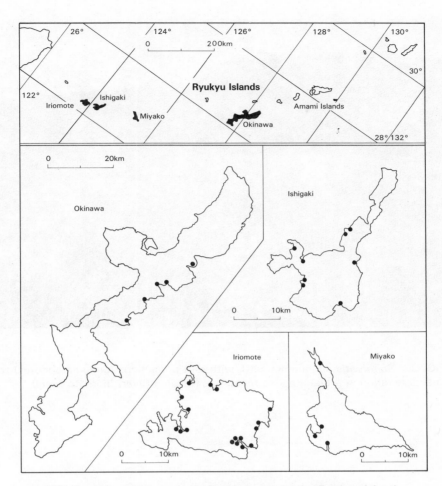

FIG. 2.   Principal locations of mangroves in the Ryukyu Islands, southern Japan

and low-growing, with five species occupying zones parallel to the coastline or the banks of tidal estuaries. This is the only island where the *Sonneratia alba* community is present (fig. 3). The *Avicennia marina* community occurs on Iriomote, Ishigaki, and Miyako, reaching its nothern limit on Miyako. Its growth is sparse and stunted. The *Kandelia candel* community also grows on Iriomote (fig. 4) and on Okinawa, reaching its northernmost limit in southern Kyushu (fig. 5). The *Rhizophora stylosa* and *Bruguiera gymnorrhiza* communities (fig. 6) are extensive on Iriomote and Ishigaki, the *Rhizophora* occupying the seaward margin, subject to deeper and more prolonged tidal submergence (fig. 7). The *Lumnitzera racemosa* community shows only stunted growth, mainly in areas where other mangroves have been removed, as on Iriomote and Ishigaki. Ecological studies of these mangrove communities by Miyawaki, Suzuki, et al. (1983) and Miyawaki, Okuda, et al. (1983) have been published.

Uses of Mangroves in Japan

Demands on mangroves to satisfy social and economic needs have not been great in Japan, but the mangroves in the Ryukyu Islands form an unusual landscape which each year attracts many Japanese tourists to the area (figs. 8–10). Secondly, mangroves help to sustain fishery as a source of food for local people (fig. 11). Finally, mangroves are being conserved for ecological studies which may help improve the productivity of this type of forest.

Mangrove vegetation in Japan has been disturbed by various human activities as described below. Some mangroves have been set aside as natural environments to be conserved, but so far this has been only partly successful. Reforestation of damaged mangrove areas on the basis of ecological and phyto-sociological research has been attempted in some of the coastal areas of southern

FIG. 3. *Sonneratia alba* community, with root projections (pneumatophores) to ensure aeration in a waterlogged habitat — Nakama River, Iriomote Island

FIG. 4. *Kandelia candel* on the southern shore of Iriomote Island

FIG. 5. *Kandelia candel* in flower at its northern limit, near Kiire, on the Satsuma Peninsula, in Kagoshima Prefecture, southern Kyushu

FIG. 6. *Bruguiera gymnorrhiza* bordering the Nakama River, Iriomote Island

FIG. 7.   *Rhizophora stylosa* bordering the Fukidori River, Ishigaki Island

FIG. 8.   Tourists visiting the mangrove area along the Nakama River, Iriomote Island

FIG. 9.   A Forestry Agency signboard for tourists, explaining mangroves
— Iriomote Island

FIG. 10.   Tourists at the largest specimen of a mangrove tree in Japan (*Heriteria littoralis*) — near the Nakama River, Iriomote Island

FIG. 11.   Catching crabs (*Scylla serrata*) in a mangrove-fringed creek near the Nakama River, Iriomote Island

Japan. Eventually, mangrove forests will have many ecological and socio-economic benefits in Japan.

Human Impact on the Mangrove Forests

Many natural environmental factors influence the species composition of mangrove communities. These factors include topography, climate, tides, salinity, strength and direction of tidal currents, and the substrate sediments, whether muddy, sandy soil, or hard ground (fig. 12). These all can be potential limiting factors in the natural vegetation of mangroves.

Human activities influence the habitats of mangrove plant communities in many ways, producing many kinds of modified vegetation. Under extreme conditions, however, the land may remain bare or support very little vegetation after the mangrove cover has been cleared. For example, if *Sonneratia alba* or *Rhizophora stylosa* communities are cut, these species recover only as scrub. If cut repeatedly, they disappear and the land remains bare.

In Japan the main forms of human impact on mangrove vegetation have been (1) cutting mangrove forests for firewood and construction material (mostly in the past), (2) disturbance by

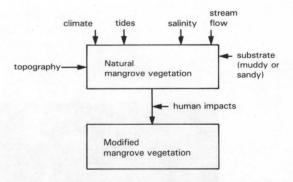

FIG. 12.   Interaction of environmental factors, including human impacts, influencing mangrove vegetation

building river-bank levees and by road construction, and (3) reclamation for urban and industrial areas. These activities alter, degrade, or completely destroy the vegetation of the mangrove community.

Conclusion

The socio-economic values of mangroves in Japan include (1) their tourist potential, (2) their fishery potential, (3) landscape protection, and (4) education for science and nature conservation. These uses are characteristic of Japanese mangrove

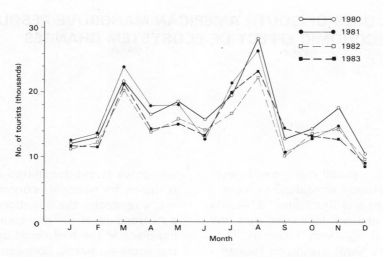

FIG. 13.   Monthly variations in the number of tourists visiting
the mangrove area on Iriomote, 1980–1983

forests. They are somewhat unusual on the world scale in that mangroves are not used by local people in ways that are common in South-East Asia (Miyawaki, Sabhasri, et al. 1983; Miyawaki 1984).

In order to make effective use of these potential values, there is a need to provide information for tourists and students about the natural ecology of mangroves. Better understanding creates more interest, and this will help to protect the natural environment and ensure that mangroves survive in the future in nature reserves which will contain potentially useful gene pools.

There are plans to make vegetation maps and guidebooks to show what kinds of mangroves there are, where they grow, and how these ecosystems function. Nature centres are being provided in special areas to educate tourists. Records of visitors to Taketono-cho on Iriomote show that tourism is greater in August than in other seasons (fig. 13). Tourists will continue to come to these areas not only to see mangroves from boats but also to explore them using wooden walkways, and so gain an understanding of the mangrove system. This must be done in such a way as to avoid putting excessive strain on the mangrove ecosystem. A successful tourist attraction is one that can be maintained to bring tourists back again and again for recreation and education.

In the Japanese perspective, mangroves are an exotic landscape, which, along with coral reefs, attract educational as well as scientific interest in the warmer southern parts of our island nation.

References

Miyawaki, A. 1984. "Phytosociological studies of mangroves in Japan and Thailand, with special reference to human impact." *Proceedings of the MAB/COMAR Regional Seminar, Tokyo*, pp. 107–110.

Miyawaki, A., ed. 1980. *Vegetation of Japan* (in Japanese), vol. 1, pp. 118–120. Shibundo Publishers, Tokyo.

——. 1981. *Vegetation of Japan* (in Japanese), vol. 2, pp. 119–121. Shibundo Publishers, Tokyo.

Miyawaki, A., S. Okuda, Y. Nakamura, S. Suzuki, Y. Murakami, K. Fujiwara, and K. Ohno. 1983. "Pflanzensoziologische Untersuchungen der Mangrove Vegetation in Japan, 2: Mangrove Vegetation auf den Inseln Okinawa, Miyako, und Ishigaki" (in Japanese with German summary). *Bulletin of the Institute of Environmental Science and Technology, Yokohama National University*, 10: 113–132.

Miyawaki, A., K. Suzuki, S. Suzuki, Y. Nakamura, Y. Murakami, Y. Tsukagoshi, and E. Nakata. 1983. "Pflanzensoziologische Untersuchungen der Mangrove Vegetation in Japan, 1: Mangrove Vegetation auf das Insel Iriomote" (in Japanese with German summary). *Bulletin of the Institute of Environmental Science and Technology, Yokohama National University*, 9: 77–89.

Miyawaki, A., Sanga Sabhasri, Sanit Aksornkoae, K. Suzuki, S. Okuda, and K. Fujiwara. 1983. "Phytosociological studies on the mangrove vegetation in Thailand, 2nd report: Mangrove vegetation of Chanthaburi and Ranong." *Bulletin of the Institute of Environmental Science and Technology, Yokohama National University*, 10: 75–111.

10. TRADITIONAL USES OF SOUTH AMERICAN MANGROVE RESOURCES AND THE SOCIO-ECONOMIC EFFECT OF ECOSYSTEM CHANGES

Samuel C. Snedaker

Up until the late 1960s, coastal mangrove forest ecosystems were considered wasteland in most parts of the world (Lugo and Snedaker 1974) and were either ignored or abused. However, in a few countries in Asia (e.g. Bangladesh, Pakistan, Malaysia, Thailand, Viet Nam) mangrove forests were viewed as natural resources that could be managed for economic gain. In contrast, the mangrove forests on the Atlantic, Pacific, and Caribbean coasts of South America were, with certain exceptions, never managed or utilized except to provide subsistence needs for local populations. Part of the reason for the benign neglect of the South American mangrove forests was the fact that most of the major population centres were located in high-altitude mountain environments (e.g. Bogota, Colombia, and Quito, Ecuador) or in areas distant from any mangrove forest (e.g. Caracas, Venezuela, and Lima, Peru). To a large extent, this preference for high, inland elevations was due to the cooler, more favourable climate (Holdridge 1967) and a relatively lower incidence of diseases such as yellow fever and malaria.

A number of authors began to document the ecological and economic values of mangroves in the 1960s and early 1970s (see Golley, Odum, and Wilson 1962; Odum 1969, 1971, Heald 1971; Snedaker and Lugo 1973). Among the cited values are the roles of mangrove forests in coastal protection (e.g. against storms and erosion), in the perpetuation of coastal water quality, and in the maintenance and the production of coastal and marine fin-fish and shellfish populations. On the basis of this new perspective, various international organizations such as Unesco, FAO, UNEP, USAID, and IUCN initiated a variety of programmes with scientific, conservation, and management objectives that resulted in widely communicated results (Rollet 1981; FAO 1982; Saenger, Hegerl, and Davie 1983; Snedaker and Snedaker 1984; Hamilton and Snedaker 1984; Snedaker and Getter 1985). In part because of a decade of international publicity, and for a variety of other reasons, most of the countries of South America now have an expanding interest in the

mangrove-forest-dominated coastal zone as a resource for national economic development. In many respects, the situation in Ecuador represents a microcosm of the changes that are rapidly taking place in the traditional uses of mangroves and the socio-economic consequences of the current coastal development. In this paper, a review of the general situation in South America is followed by a more detailed examination of a specific aspect of the situation in Educador.

The Mangrove Forests of South America

Geographic Distribution

The mangrove forests of South America extend from northern Peru on the Pacific coast to Brazil's southern state of Rio Grande do Sul on the Atlantic coast. The cold Humboldt current limits the southern extension on the Pacific coast to about 6°S latitude, whereas warm currents along the southern coast of Brazil permit limited mangrove growth to about 28°S latitude. Throughout the continent, the structural development of mangrove forests is best in areas that receive relatively high rainfall and/or abundant fresh-water runoff; a similar pattern has been documented for Central America and the Caribbean (Pool, Snedaker, and Lugo 1977). The only major exception is the delta region of the Amazon River, where there is no significant influence of salinity in the coastal zone. In the absence of salinity, mangroves cannot compete with fresh-water species which form the dominant vegetation (West 1956). In contrast, the delta of the Orinoco has extensive mangrove forests because of the seasonal periodicity in fresh-water discharge and the dry-season rise in the ambient salinity.

Mangrove Forest Area

The total area of mangrove forest land in South America, including Panama, has been estimated at some 4.6 million ha (see appendix), which represents about 22 per cent of the world's 21 million ha (Snedaker and Brown, in preparation) The national areas range from 2.5 million ha in Brazil to 2,500 ha in northern Peru:

— Brazil	2,500,000 ha
— Colombia	501,300 ha
— Ecuador	177,555 ha
— French Guiana	55,000 ha
— Guyana	80,000 ha
— Panama	486,000 ha
— Peru	2,449 ha
— Suriname	115,000 ha
— Venezuela	673,569 ha
— total	4,590,873 ha

Most of the largest single areas of undisturbed forest are found in remote areas that are largely inaccessible — for example, the Orinoco delta in eastern Venezuela (495,200 ha) and the Pacific coast of Colombia (451,300 ha). Similar expanses also occur in northern Brazil.

Species Composition

South American mangrove forests are limited to some 15 species distributed among the genera *Rhizophora, Avicennia, Laguncularia, Conocarpus,* and *Pelliciera.* This contrasts with the larger number and greater diversity of species (and genera) found in Old World mangrove forests. The diversity of associated animals is also significantly higher in the Old World, which supports the hypothesis that the centre of origin of the mangroves in the broad sense lies in the Old World.

Historical and Traditional Uses

The uses that have been made of mangroves and mangrove products are very poorly documented historically, and much of the extant information is based on conventional widom and anecdote. The earliest recorded use of mangroves is inferred from a law promulgated by King José of Portugal in 1706. The law, imposed on Brazil, made it illegal to fell mangrove trees without simultaneously utilizing the bark. It was feared that the extensive felling of trees for firewood would limit the availability of bark for the tanneries. In addition to a financial penalty the law also imposed a three-month jail term (Hamilton and Snedaker 1984).

There is little evidence in the ethnographic and archaeological literature concerning direct uses of mangroves by pre-Columbian Indians, although they are known to have inhabited coastal areas characterized by extensive mangrove forests (Meggers, Evans, and Estrada 1965). In general, pre-Columbian and historical uses are presumed to have been the same as the traditional uses that are observed today. The dominant traditional uses include the cutting of trees for firewood, charcoal, and small-diameter poles for light construction and domestic use. Each of these uses involves a small-scale operation undertaken by single families or several adults from one village. This is unlike similar activities in parts of Asia, where the harvesting and sale of poles and the production of charcoal are small to medium-size industries. In South America, and Latin America in general, the small-scale production of charcoal is inefficient and yields a product of variable quality. There is a relatively high demand for charcoal by the Panamanian middle class, and it commands a fairly high price. In Panama, the production technique is based on constructing a densely packed cone (4 m in diameter and 2–3 m high) of small logs and branch wood (25–50 cm long), covered with earth and fired from the centre. In the Panama City charcoal market, the buyers demand smokeless charcoal, which the small producers can supply only by allowing the wood to burn for an excessive period. The technique does not permit the control of the kiln temperature, and the resulting smokeless charcoal has a very low caloric value (Snedaker 1981).

Another small-scale use of mangroves has been the stripping of bark from felled *Rhizophora* trees for the production of tannin. However, the collapse of the world market for tannin has almost eliminated tannin production activities on all but a very small scale in South America. One of the larger producers is located in south-west Costa Rica and uses bark stripped from *Rhizophora* trees in Panama and illegally exported. Other than this one activity, tannin production elsewhere is presumed to be minimal and performed at the family level.

Summaries of the traditional uses of mangroves are given in Saenger, Hegerl, and Davie 1983 (by region and country) and in Hamilton and Snedaker 1984 (by species). A more detailed review of the uses of mangroves was prepared by Walsh (1977). In general, these publications confirm that much more has been reported about the traditional uses and socio-economic values of Asian than of South American mangroves. This appears to be due in part to the differences in the distribution of human populations in the two regions

Uses and Economic Value

The use of mangrove forests for economic purposes has had a long and mostly successful history in Asia. In fact, the only complete forest-management plans that exist as models for sustained forest yield are from the Asian region (e.g. Curtis 1933; Dixon 1959; Khan 1966; Choudhury 1968). No such plans exist in South America. However, there are schemes for large-scale forest utilization, albeit not on a sustainable basis.

Use of South American mangrove forests on a large, commercial scale has begun only recently, and most government-inspired efforts remain in the planning stage. At present the governments of Brazil, Panama, and Venezuela are working toward the development of forest-management plans. However, their implementation and the subsequent development of mangrove-forest-based industries is extremely slow because of the widespread opinion that mangrove wood and wood products have minimal value compared to those of other tropical forest species. One notable exception is the commercial harvesting of large *Rhizophora* trees in the Orinoco delta for use elsewhere in Venezuela as power utility poles (Hamilton and Snedaker 1984). In addition to the environmental impact (Pannier 1979), other knowledgeable observers claim that the Orinoco harvesting is exploitive and not likely to lead to an industrial base that offers permanent job opportunities in a region that is only minimally developed. Plans to use mangrove trees for timber in other countries in South America all tend to be highly exploitive rather than being based on sustained yield. In part, this is due to the short-term economics which favour the clear felling of all commercial timber at one time for sale to international wood-chip buyers. In addition to the economic situation that does not favour sustained-yield management, there is also the problem that mangrove wood and wood products are perceived as inferior to available substitutes.

Other forms of utilization of the mangrove ecosystem involve clearing the forest (with or without use of the wood) for conversion of the land to salt-evaporation ponds or maricultural ponds. Conversion to rice agriculture is not seen as an option in South America as it is in parts of Asia and Africa. Salt-evaporation ponds are limited to arid and semi-arid climates and only infrequently require the conversion of mangrove forests. Investors and developers of maricultural ponds for

the production of shrimp (mostly penaeids) also prefer semi-arid climates and seek out salt flats, barren coastal areas, and former mangrove areas for the construction of ponds. This preference is due to the fact that the land is devoid of trees, essentially flat, and close to salt water, which translates into low land-preparation and pond-construction costs. However, the extremely rapid development of the mariculture industry in South America is forcing the developers of new pond systems to convert productive mangrove forest areas as well as productive farmland (Snedaker and Dickinson, in preparation). Before 1980, only Ecuador had made a significant investment in shrimp mariculture, but the perceived financial success has inspired other countries to follow suit, sometimes with the assistance of international development organizations. As a result, most other countries have either begun encouraging mariculture or have announced plans to do so. Colombia, for example, has announced its desire to develop maricultural industries on its Caribbean and Pacific coasts.

The socio-economic impact of the widespread development of maricultural industries is only now beginning to be recognized and understood. In the following section, the experience in Ecuador is examined in greater detail because it illustrates many of the ecological, economic, and political problems of this rapidly developing industry.

Shrimp Mariculture in Ecuador: An Overview

Of the three most important Ecuadorean export crops (oil, bananas, and fishery products) shrimp from the trawler fleet and from maricultural ponds represent the top export item. In terms of foreign earnings, the shrimp industry is second only to oil. Over 70 per cent of the shrimp yield comes from managed ponds, primarily in the southern provinces of Guayas, Manabí, and El Oro, which have the greatest concentration of ponds in Ecuador. Cintron (1981a) notes that shrimp-pond yields tend to be greatest in climatic environments where the potential evapo-transpiration exceeds rainfall, which is a characteristic of southern coastal Ecuador. In 1975, shrimp production in Ecuador amounted to less than 6,000 metric tons (all sources), but by 1983 it had risen to 36,000 tons, of which some 29,000 tons were from shrimp farms (Snedaker and Dickinson, in preparation; see also Mock 1981). The rapid increase in the harvest of wild stock and the culture of shrimp in ponds throughout the world is driven by the relatively high prices that can be obtained on the world market. Ecuadorean shrimp, however, are purchased by a relatiely few large buyers for export to, and consumption in, the United States. In Ecuador, shrimp exports are said to represent an important source of foreign-currency earnings and employment opportunities along the Pacific coast.

The shrimp trawler industry had its beginnings in the 1940s, following the popularization of the Gulf of Mexico pink shrimp. The development of a shrimp fishery formed a focal industry in the Guayaquil area of Ecuador. The trawler fleet consists of day boats, which ice their catch and return to port within 24 to 36 hours, and larger vessels that refrigerate their catch and remain longer at sea. The vessels range in size from 10 to 30 m, displacing 15 to 60 tons, and are locally constructed of wood. About 20 new vessels are constructed each year. Until the mid-1970s, the trawler fleet was the dominant source of shrimp for both local consumption and the lucrative export market, and a source of local employment. The shrimp fleet still operates out of coastal Ecuador, but it is now secondary in production to the shrimp-pond industry.

Ponds for shrimp production or grow-out are intentionally located in the near-shore coastal zone, primarily for access to sea water. In Ecuador, the shrimp farms range in size from a few to several hundred hectares, with individual ponds ranging from a few hundred square meters to several hundred hectares; depths average 0.7–1.5 m (Snedaker and Dickinson, in preparation). The ponds are flooded with moderate-strength sea water, which is renewed at 10–30 per cent of volume per day. When the industry was first developing in the 1970s, the ponds were constructed on salt flats inland from the mangrove forests. However, when these habitats were no longer available, new ponds were constructed in area dominated by mangroves. In addition to expansion toward the mangrove-dominated shoreline, pond development in Ecuador is now moving to the more northern (and more humid) provinces closer to the Colombian border. As of 1985 about 60,000 ha of coastal inter-tidal and supra-tidal land has been converted to shrimp ponds. This contrasts with the 300 ha that were in production in 1970 (Snedaker and Dickinson, in preparation).

In shrimp mariculture, the ponds are stocked with postlarval (PL) shrimp for grow-out to a mature size over a period of several months. The

preferred species are *Penaeus vannamei* and, to a lesser extent, *P. stylirostris* (Cun 1982a), which are stocked at rates of 50,000–250,000 PLs per hectare and higher (Cun 1982b). Following a five- to seven-month grow-out period, the pond yields range from 180 to 360 kg/ha/yr, although feed and other subsidies can double or triple the yields (Mock 1981). Few pond owners use feed supplements, believing that the marginal gain does not justify the cost. Penaeid PLs are obtained by netting wild shrimp populations that are present in local estuarine waters. The gathering or harvesting of PLs is itself a substantial industry that provides much significant employment. The netting of PLs is done by individuals, who are each able to collect 20,000–50,000 PLs during an outgoing tide. The individual harvests are sold to on-site brokers, who in turn supply the pond-owners/managers (Snedaker and Dickinson, in preparation).

In the late 1970s and early 1980s, numerous complaints began to be sounded over the decline or scarcity of PLs. Complaints reached a peak prior to the 1982/83 "El Niño" event. The problem has continued, as can be seen from the fact that the area in production ponds has increased by 60 per cent just since 1982, while the annual harvest of PLs has remained more or less constant at less than 10 million animals (Snedaker and Dickinson, in preparation). The common opinion in Ecuador is that the destruction of coastal mangrove forests, the nursery grounds for many species of shrimp (as well as fin fish and shellfish), is the cause of the decline in PLs (Fundación Natura 1981). The concern appears to be valid, based on the results of quantitative studies (Turner 1977; Martosubroto and Naamin 1977) which independently showed that harvested yields of commercial shrimp are proportional to the total area in coastal mangroves and marshes. Turner's data (1977) show an average yield of 767 kg of shrimp per hecatare of mangrove forest per year (about twice the annual yield from minimally managed grow-out ponds). Thus, the loss of mangrove and associated inter-tidal area can be presumed to have a significant effect on local shrimp populations. Other possible causes for the decline include abnormal sea-surface temperatures and salinity, and near-shore water pollution (Snedaker and Dickinson, in preparation).

As a result of the unpredictable availability of PLs, there is now considerable interest in Ecuador in establishing one or more hatcheries for their production (Instituto Nacional de Pesca 1982). Ostensibly, this would relieve the pressure on the wild stocks and guarantee continuing supplies of seed shrimp to the pond producers. However, this approach would favour the large producers who can provide the capital investment and operating funds for a hatchery while forcing small and marginal producers to continue to rely on wild PLs, stocks. Should hatchery production techniques succeed in supplying the entire demand for PLs, the economic incentive to the industry for preserving the mangrove source of PLs would be lost.

Because of the interest in promoting the pond culture of shrimp while preserving the mangrove habitat both for the perpetuation of shrimp and for other reasons, significant technical research has been undertaken in Ecuador, particularly in the Guayaquil area. There are many useful papers on techniques relating to pond production (e.g. Cun 1982a, 1982b; Cun and Marin 1982; Cun and Regalado 1982; Yoong and Reinoso 1982). The interest in pond production is also matched by a widespread interest in the mangrove ecosystem as an important natural resource (e.g. Cintron 1981a, 1981b; Cintron and Schaeffer-Novelli 1981a; Horna, Medina, and Macias 1980; Valverde 1980), mainly because of the relationship with shrimp. In addition, there are a variety of published recommendations offering alternatives to general problems that affect the industry (cf. Barniol n.d.; Hallberg 1977; Primer Congreso Nacional de Productores de Camarón 1982). However, with the exception of a US Agency for International Development research study currently being completed by Snedaker and Dickinson (in preparation), there has been no socio-economic analysis of the shrimp industry per se, nor has there been an evaluation of the effect that the development of the industry has had on the life-styles and well-being of the local people.

Many problems affecting the shrimp industry came into focus during the massive flooding in the coastal zone that was associated with the 1982/83 El Niño phenomenon, whose severity brought world-wide disaster relief to Ecuador. The significance of its recurrence resulted in pressures on the government to undertake major infrastructural investments in flood control and drainage. These actions, coupled with hydro-electric and irrigation projects being implemented or contemplated, added more pressures on coastal ecosystems, natural and artificial, that depend on seasonal pulses of fresh water and sediment. Erosion is already a serious problem in the Machala area, where there is a high density of producing ponds. In spite of the 1982/83 El Niño event and the fo-

cus of attention on the coastal zone, there are still no effective plans for managing the shrimp-farm industry for its own benefit and the economic benefit of the region.

The expansion of the area in shrimp grow-out ponds in Ecuador, in the absence of a guiding policy and management protocol, has resulted in a number of development problems. Some of the more significant or well-publicized problems include (1) a significant reduction in mangrove area and interruption in drainage patterns, which affect the availability of shellfish and fin fish in off-shore and estuarine waters for local and national consumption, (2) the loss of the mangrove forest as a source of wood products, (3) the loss of mangrove nursery area for the postlarval shrimp used in grow-out ponds, (4) highly variable and generally low shrimp-pond production because of a lack of technical know-how, (5) salinization of valuable irrigated farmland in many provinces caused by shrimp ponds located on farmland, (6) leaching and drainage of pesticides and herbicides from active farmlands into the near-shore waters, and (7) potential reduction in shrimp-pond and mangrove production alike due to upstream infrastructures causing fresh-water and sediment starvation.

Among scientists who have studied the Ecuadorean shrimp-pond industry, there is a pervasive feeling that the socio-economic problems may far outweigh the technical and/or management problems (Snedaker and Dickinson, in preparation). It is also felt that any solutions to the latter problems cannot be effectively implemented without a prior resolution of the socio-economic problems. Among the foremost issues are (1) the concentration of shrimp-pond wealth and knowledge in the hands of a relatively few entrepreneurs coupled with a perceptible shift to foreign ownership, (2) the inevitable decline in job opportunities as pond contruction peaks out and hatcheries substitute for the extensive harvesting of wild PLs, (3) the "flight" of maricultural earnings to foreign banks, and (4) smuggling across the Colombian and Peruvian borders to acquire PLs, to avoid Ecuadorean export taxes, and to make profits on discrepancies in currency exchange rates.

Acknowledgements

This paper was developed from my experiences in Latin American dating back to the mid-1960s. Some of the specific data and information on shrimp mariculture in Ecuador were taken from in-country research supported by the US Agency for International Development, Program in Science and Technology Cooperation, Grant No. DPE-5542-G-SS-4022-00. The paper was typed and assembled by Tropic House International, Inc., of Miami, Florida, USA.

Appendix: Mangrove Forest Area in South America, by Country and Locality

	Area (hectares)
Brazil	2,500,000[a]
Amapa	250,000[b]
Espírito Santo	30,000[c]
Pará	400,000[d]
Piauí	47,700[a]
Maranhão	602,300[e]
São Luís	226,000[e]
Tutóia	26,000[e]
Turiaçu	207,300[e]
Cururupu	100,000[e]
Delta do Parnaíba	24,000[e]
Itapecuru	19,000[e]
Colombia	501,300[f]
Caribbean coast	50,000[f]
Pacific coast	451,300[f]
Ecuador	177,555[g]
Esmeraldas	40,300[h]
Manabí	6,000[i]
Guayas	90,190[j]
El Oro	40,265[j]
Galapagos	800[j]
French Guiana	55,000[k]
Guyana	80,000[l]
County of Berbice	30,000[l]
County of Demerara	10,000[l]
County of Essequibo	40,000[l]
Panama	486,000[m]
Darién	70,430[n]
Panamá	33,550[n]
Peru	2,449[o]
Boca Capones	1,358[o]
Matapala	390[o]
El Gallo	63[o]
Mantanza	124[o]
Chinchona	10[o]
Zarumilla	480[o]
Algarrobo	60[o]
Gallegos	20[o]
Soledad	25[o]
Juanita	50[o]
Envida	136[o]
Boca del Bendito	385[o]
Bahía Puerto Pizarro	134[o]
Jely	71[o]
Puerto Rico	21[o]
La Gianina	42[o]
Boca del Estero Hondo	328[o]
Tamarindo	38[o]
Alamo	40[o]
Estero Hondo	250[o]
Bocana del Río Tumbes	140[o]
Estero Correles	140[o]
Suriname	115,000[p]
Venezuela	673,569[q]
Western region	15,468[q]
Central-western region	15,616[q]
Central region	6,608[q]
Central-eastern region	138,377[q]
Orinoco delta	495,200[q]
Margarita island	2,300[q]

Note: This appendix has been modified from Snedaker and Brown (in preparation).

Brazil

a. Data obtained from FAO/PNUMA 1981, cited in Cintron and Schaeffer-Novelli 1981b. Although there are a number of estimates of the mangrove forest area in Brazil, the FAO/PNUMA report is reported to be the best estimate (Gilberto Cintron and Yara Schaeffer-Novelli, personal communication).

b. Data obtained from Hueck 1966, cited in Weishaupl 1981.

c. Data obtained from Ruschi 1950, cited in Weishaupl 1981.

d. Data obtained from Brazil 1974a and 1974b, cited in Weishaupl 1981.

e. Data obtained from Instituto de Recursos Naturais do Estado do Maranhão 1975.

Colombia

f. Data provided by Jorge Hernán Torres Romero (personal communication); they are in general agreement with the majority of other estimates obtained from Colombia, including Colombia 1967. FAO/PNUMA 1981 gives a total area of 450,000 ha, of which 287,000 ha are located on the Pacific coast. There is an extensive area of tropical lowland forest on the Pacific coast which grades into the mangrove forest. It is possible that the higher estimates are the result of an inadequate differentiation between the two forest types. However, I believe that the 287,000 ha estimate for the Pacific coast may be too conservative.

Ecuador

g. Total area estimate obtained by summarizing the individual areas. FAO/PNUMA 1981 estimates the total area at 235,000 ha.

h. There are various estimates for Esmeraldas, including 8,000 ha (Berthon 1959), 29,600 ha (Acosta-Solís 1957), 40,300 ha (Dixon et al., n.d.) and 180,800 ha (Rafael R. Horna Zapata, Francisco Yoong Basurto, and Blanca Reinoso de Ayeiga, personal communication). The estimate from Dixon et al. is considered to be a reasonable provisional value and is used in the summary.

i. Data provided by Rafael R. Horna Zapata, Francisco Yoong

Basurto, and Blanca Reinoso de Ayeiga (personal communication). Ministerio de Agricultura y Ganadería 1980 gives an estimate of 14,700 ha, but this is believed to be too high (Gilberto Cintron, personal communication).

Data obtained from Cintron (1981a, 1981c). In addition to the mangrove areas, there are salt flats comprising 42,712 ha in Guayas and 13,024 in El Oro. These areas are not included in the estimate used.

A large variety of estimates for the mangrove area in Ecuador were obtained, and the more conservative alternatives are used in the report. For example, Dixon et al. (n.d.) state that there are 403 km² of mangrove forest in the San Lorenzo-Limones area, of which 290 km² are in regenerating forest and an additional 113 km² in ''degraded'' forest (Gilberto Cintron, personal communication). Other estimates for Ecuador, provided by Rafael R. Horna Zapata, Francisco Yoong Basurto, and Blanca Reinoso de Ayeiga (personal communication), may be unrealistically high, particularly for Esmeraldas:

Ecuador	316,800 ha
Esmeraldas	180,000 ha
Manabí	6,000 ha
Guayas	80,000 ha
El Oro	50,000 ha
Galapagos	800 ha

French Guiana
k. Data obtained from FAO/PNUMA 1981.
Guyana
l. Data provided by C. A. Persaud and Reuben Charles (personal communication). FAO/PNUMA 1981 gives an estimate of 150,000 ha, but the more conservative value is probably the best estimate.
Panama
m. Data obtained from FAO/PNUMA 1981, cited by G. Cintron (personal communication). Cintron also cites FAO/PNUD 1972 as giving a total mangrove area for Panama of 409,210 ha. One possible reason for the large variation in estimates may be a difference in whether or not contiguous lowland forests, such as those dominated by cativo (*Prioria copaifera*) and orey (*Campnosperma panamensis*), are included in the estimates. The range of estimates for Panama include the following: 104,000 ha (Donaldson et al. 1963), 199,000 ha (Falla 1978a), 505,600 ha (Falla 1978b).
n. Data (in acres) from Donaldson et al. 1963 obtained by G. Cintron.

The actual distribution of mangrove areas among the other provinces of Panama is also questionable. I obtained data from the offices of the Ministerio de Desarrollo Agropecuario, Dirección Nacional de Recursos Naturales Renovables, which are probably representative of the *relative* distribution among the provinces:

Panama	297,532 ha
Bocas del Toro	64,010 ha
Cocle	25,125 ha
Chiriqui	66,645 ha
Darien	28,225 ha
Herrera	8,450 ha
Los Santos	8,800 ha
Panamá	122,925 ha

Peru
o. Data (in square meters) provided by Miguel A. Checa L. (personal communication). This estimate is an order of magnitude lower than the 28,000 ha estimated by FAO/PNUMA 1981.
Suriname
p. Data obtained from FAO/PNUMA 1981.

Venezuela
q. Data provided by F. Pannier (personal communication). FAO/PNUMA 1981 gives a total area for Venezuela of 260,000 ha, but it is believed that this reflects only the extensive areas of potentially commercial forest.

References

In addition to literature cited, the following list includes the sources of personal communications.

Acosta-Solís, M. 1957. *Los manglares del Ecuador.* Contribution no. 29. Instituto Ecuatoriano de Ciencias Naturales, Quito, Ecuador.

Barniol Zerega, R. n.d. *Diagnóstico y recomendaciones sobre el recurso camarón.* Subsecretaría de Recursos Pesqueros, Quito, Ecuador.

Berthon, P. F. 1959. *Informe al gobierno del Ecuador sobre el desarrollo de las industrias forestales en las regiones de Guayaquil y San Lorenzo.* Report no. 1125. Food and Agriculture Organization/ETAP, Rome.

Brazil. 1974a. Projeto RADAM. Folha SA 22, Belém. Vol. 5. Departamento Nacional da Produção Mineral, Rio de Janeiro. Cited in Weishaupl 1981.

——. 1974b. Projeto RADAM. Folha NA/NB 22, Macapá. Vol. 6. Departamento Nacional da Produção Mineral, Rio de Janeiro. Cited in Weishaupl 1981.

Charles, Reuben. Personal communication. Principal Fisheries Officer. Fisheries Division, Ministry of Agriculture, 39 Brickdam, Stabroek, Georgetown, Guyana.

Checa L., Miguel A. Personal communication. Jefe de investigación y desarrollo. Tecnológica de Acuacultura y Pesca S.R.L., Av. Salaverry 2447, Lima 27, Peru.

Choudhury, A. M. 1968. *Working plan of Sundarban Forest Division for the period from 1960–61 to 1979–80.* Vol. 1. East Pakistan Government Press, Tejagon, Dacca.

Cintron, G. 1981a. *El manglar de la costa ecuatoriana.* Departamento de Recursos Naturales, Puerto Rico.

——. 1981b. ''Bibliografía sobre los manglares de Ecuador.'' Seminario sobre ordenación y desarrollo integral de las zonas costeras, 18–27 May 1981, Guayaquil, Ecuador.

——. 1981c. *Los manglares de Santa Catarina.* Oficina Regional de Ciencia y Tecnología para América Latina y el Caribe de Unesco. Universidade Federal de Santa Catarina, Florianópolis, Brazil.

Cintron, Gilberto. Personal communication. Departamento de Recursos Naturales, Box 5887, Puerta de Tierra, Puerto Rico 00906.

Cintron, G., and Y. Schaeffer-Novelli. 1981a. ''Introducción a la ecología del manglar.'' Seminario sobre ordenación y desarrollo integral de las zonas costeras, 18–27 May 1981, Guayaquil, Ecuador.

——. 1981b. *Los manglares de la costa Brasileña: Revisión preliminar de la literatura.* Oficina Regional de Ciencia y Tecnología para América Latina y el Caribe de Unesco. Universidade Federal de Santa Catarina, Florianópolis, Brazil.

Colombia. 1967. *Mapa general de bosques.* Vol. 3, no. 2. Instituto Geografico ''Augustín Codazzi,'' Departamento Agricola, Bogota.

Cun, M. 1982a. ''Guía práctica para la cría de camarones comerciales (*Penaeus*) en Ecuador.'' *Boletín Científico y Técnico* (Instituto Nacional de Pesca, Guayaquil, Ecuador), vol. 5, no. 1.

——. 1982b. ''Especies de camarones marinos (*Penaeus*) que se han adaptado a las condiciones de cultivo en Ecuador.''

Boletín Científico y Técnico, vol. 5, no. 3.

Cun, M., and C. Marin. 1982. ''Estudio de los desembarques del camarón (gen. *Penaeus*) en el Golfo de Guayaquil (1965–1979).'' *Boletín Científico y Técnico,* vol. 5, no. 3.

Cun, M., and M. Regalado. 1982. ''Experiencias en laboratorio con alimentación suplementaria utilizada para camarones marinos (*Penaeus vannamei* y *P. stylirostris*).'' *Boletín Científico y Técnico,* vol. 5, no. 3.

Curtis, S. J. 1933. *Working plan for the Sundarbans Division (1931–51).* Forest Department, Bengal.

Dixon, R. G. 1959. *A working plan for the Matang Mangrove Forest Reserve, Perak.* Perak State Forestry Department Publication. Perak, Malaya.

Dixon, R. G., J. W. Eastwood, O. P. Ortiz, and G. Gortaire. n.d. *Tipos de bosque en la zona del proyecto del programa de desarrollo forestal de noroccidente.* United Nations Special Fund Project no. 127, Quito, Ecuador.

Donaldson, P. O., K. M. Macdonald, J. W. Rankin, and T. A. McKenzie. 1963. *Forests of the Republic of Panama: Resource development, industrial development, product potential.* Greenacres, Inc., Consulting Foresters, 4501 Rainier Ave. S., Seattle, Wash., USA.

Falla, A. 1978a. *Plan de desarrollo forestal: Estudio de las perspectivas del desarrollo forestal en Panamá.* Informe técnico No. 2. Food and Agriculture Organization, Panama. (FAO/PCT/6/PAN/02/I)

——— . 1978b. *Plan de desarrollo forestal: Estado actual del subsector.* Informe técnico No. 1. Food and Agriculture Organization, Panama. (FAO/PCT/6/PAN/01/I)

FAO. 1982. *Management and utilization of mangroves in Asia and the Pacific.* Environment paper no. 3. Food and Agriculture Organization, Rome.

FAO/PNUD. 1972. *Panamá: Reconocimiento general de los bosques e inventario detallado de azuero. III, Resultados y comentarios.* Informe técnico 12. (FO:SF/PAN 6)

FAO/PNUMA. 1981. *Proyecto de evaluación de los recursos forestales tropicales.* Informe técnico 1. (UN 32/6.1301-78—04)

Fundación Natura. 1981. *Diagnóstico de la situación del medio ambiente en el Ecuador.* Quito, Ecuador.

Golley, F. B., H. T. Odum, and R. F. Wilson. 1962. ''The structure and metabolism of a Puerto Rican red mangrove forest in May.'' *Ecology,* 43: 9–19.

Hallberg, R. O. 1977. *Research programme for the marine environment of the Gulf of Guayaquil, el Ecuador.* Consultative Panel on the Coastal Zone, 27–29 Aug. 1979. Unesco, Paris.

Hamilton, L. S., and S. C. Snedaker. 1984. *Handbook for mangrove area management.* United Nations Environment Programme and East-West Center, Honolulu, Hawaii, USA.

Heald, E. J. 1971. *The production of organic detritus in a south Florida estuary.* Sea Grant Technical Bulletin no. 6. University of Miami, Coral Gables, Fla., USA.

Holdridge, L. R. 1967. *Life zone ecology.* 2nd ed. Tropical Science Center, San José, Costa Rica.

Horna Zapata, Rafael R. Personal communication. Escuela Superior Politécnica del Litoral, Rocafuerte y Loja, Apartado 5863, Guayaquil, Ecuador.

Horna, R., F. Medina, and R. Macias. 1980. *Breve estudio sobre el ecosistema del manglar en la costa Ecuatoriana.* Escuela Superior Politécnica del Litoral. Guayaquil, Ecuador.

Hueck, K. 1966. *Die Wulder Sudameerikas: Ökologie, Zusammensetzung und wirtschaftliche Bedeutung.* Gustav Fischer, Stuttgart, Cited in Weishaupl 1981.

Instituto de Recursos Naturais do Estado do Maranhão. 1975. *Mangue do Maranhão: Estudo em Alcantara.* COTEC, São Luís, Brazil.

——— . 1976. *Mangue-incorporação a economia maranhense.* Pp. 1–12. COTEC, São Luís, Brazil. Cited in Weishaupl 1981.

Instituto Nacional de Pesca. 1982. *Proyecto: Estación tecnológica para cultivo del camarón.* Guayaquil, Ecuador.

Khan, S. A. 1966. *Working plan of the Coastal Zone Afforestation Division from 1963–64 to 1982–83.* Agriculture Department, Government of West Pakistan, Lahore.

Lugo, A. E., and S. C. Snedaker. 1974. ''The ecology of mangroves.'' *Annual Review of Ecology and Systematics,* 5: 39–64.

Martosubroto, P., and N. Naamin. 1977. ''Relationship between tidal forests (mangroves) and commercial shrimp production in Indonesia.'' *Marine Research in Indonesia,* 18: 81–86.

Meggers, Betty J., Clifford Evans, and Emilio Estrada. 1965. *Early formative period of coastal Ecuador: The Valdivia and Machalilla phases.* Smithsonian Institution, Washington, D.C.

Ministerio de Agricultura y Ganadería. 1980. *Diagnóstico regional de Manabí (Resumen).* Centro de Rehabilitación de Manabí. Departamento de Planificación Regional. Editorial Gregorio (Noviembre), Puerto Viejo, Ecuador.

Mock, C. R. 1981. ''Report on penaeid shrimp culture consultation and visit — Guayaquil, Ecuador, South America, and Panama, Central America, August 12 to September 20, 1981.'' Southeast Fisheries Center, Galveston Laboratory, Galveston, Tex., USA.

Odum, W. E. 1969. ''The structure of detritus-based food chains in a south Florida mangrove system.'' Ph.D. dissertation. University of Miami, Coral Gables, Fla., USA.

——— . 1971. *Pathways of energy flow in a south Florida estuary.* Sea Grant Technical Bulletin no. 7. University of Miami, Coral Gables, Fla., USA.

Pannier, F. 1979. ''Mangrove impacted by human-induced disturbances: A case study of the Orinoco delta mangrove ecosystem.'' *Environmental Management,* 3 (3): 205–216.

Pannier, Federico. Personal communication. Professor of Plant Physioecology. Facultad de Ciencias, Departamento de Botánica, Universidad Central de Venezuela, Apartado 80.390, Caracas 1080-A, Venezuela.

Persaud, C. A. Personal communication. Guyana Forestry Commission, Water Street, Kinston, Georgetown, Guyana.

Pool, D. J., S. C. Snedaker, and A. E. Lugo. 1977. ''Structure of mangrove forests in Florida, Puerto Rico, Mexico and Costa Rica.'' *Biotropica,* 9 (3): 195–212.

Primer Congreso Nacional de Productores de Camarón. 1982 *Recomendaciones de la Comisión Científico-Técnia.* Guayaquil, Ecuador.

Reinoso de Ayeiga, Blanca. Personal communication. Instituto Nacional de Pesca, Casilla 5918, Guayaquil, Ecuador.

Rollet, B. 1981. *Bibliography on mangrove research, 1600–1975.* Unesco, Paris.

Ruschi, A. 1950. ''Fitogeografia do estado do Esp. Santo — I: Considerações gerais sobre a distribuição da flora no estado do E. Santo.'' *Boletim do Museu de Biologia Prof. Mello-Leitao,* série botânica, 1: 1–353. Santa Teresa, Espírito Santo, Brazil. Cited in Weishaupl 1981.

Saenger, P., E. J. Hegerl, and J. D. S. Davie, eds. 1983. *Global status of mangrove ecosystems.* Commission on Ecology Papers, no. 3. International Union for Conservation of Nature and Natural Resources, Gland, Switzerland. (*The Environmentalist,* vol. 3, supplement no. 3).

Schaeffer-Novelli, Yara. Personal communication. Instituto Oceanografico da Universidade de São Paulo, Cidade Universitaria C.P. 9075, CEP 05508, São Paulo, Brazil.

Snedaker, S. C. 1981. *Mangrove forest management and utili-*

zation in Panama. FAO Forest Industries Development Project in Latin America. (RLA/77/019)

Snedaker, S. C., and M. S. Brown. In preparation. *Biosphere inventory of mangrove forest lands: Total area, current status, managing institutions and research initiatives.* United States Forest Service/United States Man and Biosphere Program, Washington, D.C.

Snedaker, S. C., and J. C. Dickinson. In preparation. "Shrimp pond siting and mangement alternatives in Ecuador, with appendices by M. S. Brown and E. Lahmann." Final report to the United States Agency for International Development on Grant Number DPE-5542-G-SS-4022-00. University of Miami, Coral Gables, Fla., USA.

Snedaker, S. C., and C. D. Getter. 1985. *Coasts: Coastal resources management guidelines.* Coastal Publication no. 2, Renewable Resources Information Series. US Agency for International Development, and National Park Service, US Department of Interior, Washington, D.C.

Snedaker, S. C., and A. E. Lugo. 1973. *The role of mangrove ecosystems in the maintenance of environmental quality and a high productivity of desirable fisheries.* Contract no. 14-16-008-606. Bureau of Sports Fisheries and Wildlife, Atlanta, Ga., USA.

Snedaker, S. C., and J. G. Snedaker, eds. 1984. *The mangrove ecosystem: Research methods.* Unesco, Paris.

Torres Romero, Jorge Hernán. Personal communication. Profesor. Universidad Nacional de Colombia, Calle 63-A, No. 21-42, Bogotá, D.E., Colombia.

Turner, R. E. 1977. "Intertidal vegetation and commercial yields of penaeid shrimp." *Transactions of the American Fisheries Society,* 106 (5): 411–416.

Valverde, F. de María. 1980. "Proyecto: Estudio del ecosistema de manglares en el Ecuador." Proposal submitted to Unesco, Paris. Facultad de Ciencias Naturales, Universidad de Guayaquil, Ecuador.

Walsh, G. E. 1977. "Exploitation of mangal." In V. J. Chapman, ed., *Ecosystems of the World.* Vol. 1, *Wet coastal ecosystems,* pp. 347–362. Elsevier Scientific Publishing Company, New York.

Weishaupl, Liane. 1981. "Plantas de mangue: Contribuição ao conhecimento de *Laguncularia racemosa* (L.) Gaertn. f. (Combretaceae)." Sección, "Distribución geográfica de los manglares brasileiros y bibliografia correspondiente (parcial)." Master's thesis. Instituto de Biociencias da Universidade de São Paulo, São Paulo, Brazil.

West, R. C. 1956. "Mangrove swamps of the Pacific coast of Colombia." *Association of American Geographers Annals,* 46: 98–121.

Yoong Basurto, Francisco. Personal communication. Doctor Médico, Veterinario. Instituto Nacional de Pesca, Casilla 5918, Guayaquil, Ecuador.

Yoong Basurto, F., and B. Reinoso Naranjo. 1982. "Cultivo del camarón marino (*Penaeus*) en el Ecuador: Metodologías y técnicas utilizadas. Recomendaciones." *Boletín Científico y Técnico* (Instituto Nacional de Pesca, Guayaquil, Ecuador), vol. 5, no. 2.

RECOMMENDATIONS WITH RESPECT TO THE SPECIAL CASE OF THE MANGROVE FOREST OF THAILAND

Summarized by Eric Bird and Peter Kunstadter

The following recommendations were approved at the conclusion of the Workshop on the Socio-economic Situation of Human Settlements in Mangrove Forests.

A. Philosophy

An underlying philosophy for a mangrove land-use zoning plan was recommended:

1. Mangroves are valuable resources, even in their natural state, and not wasteland. They should be preserved as much as possible for their multiple values as sources of economic, ecological, scientific, and cultural benefits now and for future generations. Because approximately 30 per cent of Thailand's mangrove forests have already been destroyed by conversion to other uses or by pollution, further conversion or destruction should be reduced to a minimum.

2. Plans for the use of mangrove areas should emphasize sustained productivity through multiple use rather than conversion for a single, exclusive use.

3. Plans for mangrove land-use zoning should be co-ordinated with national socio-economic development plans, both in terms of designated regions for industrial development and with regard to the socio-economic objectives of increasing the equity of distribution of development, especially for the benefit of the relatively poor rural majority of Thailand's population.

4. Research in Thailand and elsewhere has repeatedly demonstrated that changes within a river catchment (e.g. deforestation, cultivation and associated soil erosion and pesticide contamination, mining, modification of water flow and drainage) or in the adjacent marine environment (e.g. pollution, dredging, modification of tidal flow) can modify or destroy mangrove ecosystems. Likewise it has been demonstrated that change within the mangroves may have serious conse-

quences on upstream and marine fisheries and on protection of the coastline. It is therefore recommended that planning for mangrove land-use zoning should be integrated with planning for coastal development and that upstream development plans should be examined for their potential effects on the mangroves.

B. Data to Be Collected

The following types of data required for rational scientific planning of mangrove land-use zones should be collected.

1. The general pattern of distribution of mangrove forests in Thailand is well known, but more detailed information is required concerning the distribution and quality of forest, landforms, and detailed information is required concerning the distribution and quality of forest, landforms, and clude information on current patterns of use and on the dynamics of existing mangrove areas (e.g. areas of die-back and erosion or of sedimentation and accretion). Information should be tabulated on the environmental factors that sustain or constrain the mangrove ecosystem (patterns and rates of flow of fresh and brackish water, salinity levels, sediment gains and losses, nutrient budgets, sources of pollution) and on the geo-morphological and ecological features associated with mangrove expansion or contraction.

2. Quantities and direct economic values of mangrove forest and fisheries resources are fairly well known. The distribution of the benefits of these resources (e.g. as used for subsistence by people living in and near these forests, or the basis for employment of people in and around the mangrove area — e.g. in transporting, processing, and selling mangrove products) and the value of ecological effects (e.g. coastline protection, breeding grounds for fish caught in the mangrove area and elsewhere) have not been estimated. The distribution of economic benefits and costs of replacement of the environmental functions of the mangrove-forest ecosystem should be estimated

113

and included in cost/risk/benefit calculations of alternative plans for management or development.

3. The socio-economic condition of mangrove residents is known from only a few studies, which were presented in the workshop. More extensive information is needed concerning the numbers of people living in, and directly or indirectly dependent on, mangrove resources. This information should include estimates of demographic rates and their determinants, including family planning, in- and out-migration, and the origins, destinations, and motivations of migrants, and should also include estimates of the level and distribution of income and access to infrastructural services among dwellers in the mangrove zone.

4. Marketing patterns should be studied, including access to markets and organization and terms of trade (e.g. distance and transportation to markets, trade through middlemen or entrepreneurs, level of return on investment, interest charges) to provide a basis for assessing who will benefit from specific proposed development in and around the mangrove areas.

C. Specific Recommendations

The following specific recommendations are made for developing a plan for zoning land use in mangrove areas, making use of the philosophy and data outlined above.

1. Zones should be outlined for the following types of activities:

(a) Conservation, including the protection of natural and relatively undisturbed mangrove ecosystems as samples of those occurring in Thailand. The conservation zones will maintain species and genetic diversity and provide areas for scientific research and for education, recreation, and cultural interest. At the same time they will provide coastline protection and breeding grounds and shelter for fish and shellfish. Conservation zones should be declared mangrove reserves and managed appropriately. Mangrove areas designated for educational purposes should be located as near as possible to centres of population and transportation to ensure that they are well used. Additional reserves should be designated in order to preserve genetic diversity and seed stock, including the full range of environmental zonation required for sustained regeneration of the mangrove-forest types represented in Thailand.

(b) Management for sustained yield, primarily for timber production, using harvesting and reforestation methods that minimize environmental impact. Where appropriate, this may be accomplished while at the same time increasing the speed of regeneration and the proportion of timber trees. Concurrently, such areas will serve as breeding grounds and shelter for fish and shellfish and will continue to provide shoreline protection.

(c) Management for sustained yield, primarily for fisheries, maintaining a habitat that sustains the population of fish and crustaceans that can be harvested in the mangrove area and in adjacent estuarine, lagoonal, or marine waters. Such areas will continue to provide areas for growth or regrowth of forest species and for coastline protection, as well as to serve as breeding grounds and shelters for fish and shellfish. In practice it should often be possible to combine management for both fisheries and forestry.

(d) Conversion of mangrove areas for other uses, such as aquaculture (fish and shrimp ponds), salt farms, agriculture, urban or industrial development. All these uses require clearing and destruction of mangrove ecosystems. Because of the destructive nature of this form of land use, zoning for this purpose should be kept to a minimum, preferably on sites that have already been converted. Because extensive conversion of mangroves to shrimp or fish ponds destroys breeding grounds for larvae, which are an essential resource for the continued productivity of these farms, coastal zone planning should attempt to locate such activities inland from the coastline. As much as possible, mangroves should be retained for their multiple uses as outlined above.

(e) Waste disposal. Use of mangrove areas for disposal of urban and industrial waste and overburden from dredging and mining should be discouraged. Where waste-disposal and mining activities occur in mangrove areas, the damaged mangrove vegetation and natural drainage patterns should be restored when the mining is completed. Costs of reforestation should be included in cost/benefit calculations of mining operations.

(f) Reforestation. Reforestation should be planned where mangroves have been badly damaged by waste disposal, including reconstruction of appropriate drainage and replanting. Research in Thailand has already demonstrated the feasibility of mangrove reforestation.

2.   An appropriate schedule of fees or taxes should be charged in order to discourage conversion of mangrove areas into areas exclusively for fish or shrimp ponds, salt farms, or waste dumps. Funds collected from this source should be used to finance the reforestation of mangroves in previously converted areas that have been abandoned. Taxes will also have the effects of assigning the costs of development directly to the beneficiaries and discouraging development for single, exclusive uses.

3.   Plans for land-use zoning in the mangrove areas should take account of the socio-economic condition of people who live in settlements within or close to the mangrove area and who are wholly or partly dependent on mangrove resources for their livelihood. Areas in which there are already villages may be allocated for forestry or fisheries development, but should ordinarily not be allocated for conservation, so as to avoid conflicts between the villagers and government officials.

4.   To facilitate effective planning, laws and regulations should be revised to deal with the specific conditions and issues of mangrove management and to eliminate gaps and conflicts in existing legislation and regulations.

5.   Laws and regulations should be backed by an enforcement mechanism, with sufficient officers trained for enforcement, together with the necessary support equipment, including vehicles and boats, to enable them to carry out their responsibilities.

6.   Widespread education is necessary to ensure public support for legislation and to enforce regulations controlling land use in mangrove areas. Targets for special educational efforts include those who live in and near mangrove areas and public officials, administrators, and legislators whose work concerns mangrove areas. Methods for education should include: the integration of examples from mangrove ecology into primary- and secondary-school biology courses, the training and use of extension officers for mangrove areas, the presentation on radio or television of information for people who live in or near the mangroves, and the aforementioned establishment of mangrove educational areas in the form of botanical gardens or conservation zones accessible from centres of population. Educational programmes should emphasize the ecological and economic value of mangrove ecosystems as a national resource and should help to generate support for, compliance with, and enforcement of regulations protecting the mangroves.

7.   Specific steps should be taken to ensure that residents of mangrove-area villages benefit from developments that take place in these areas. For example, where surveys indicate it is feasible, adjacent waters should be used for shellfish or caged fish culture to increase subsistence or cash income for these people.

WORKSHOP PARTICIPANTS

Dr. Sanit Aksornkoae
Faculty of Forestry
Kasetsart University
Bangkhen, Bangkok 10900
Thailand

Dr. Eric Bird
Department of Geography
University of Melbourne
Parkville, Victoria 3052
Australia

Miss Jenjai Boonumpai
Faculty of Forestry
Kasetsart University
Bangkhen, Bangkok 10900
Thailand

Dr. A. C. J. Burgers
Consultant
Development Studies Division
The United Nations University
Toho Seimei Building
15–1, Shibuya 2-chome
Shibuya-ku, Tokyo 150
Japan

Dr. Chan Hung Tuck
Forest Ecologist
Forest Research Institute
Kepong, Selangor
Malaysia

Mrs. Sriprai Chaturongakul
Research Project and Co-ordination Division
National Research Council
Bangkhen, Bangkok 10900
Thailand

Mr. Vipak Jintana
Royal Forest Department
Bangkhen, Bangkok 10900
Thailand

Mr. Jitt Kongsangchai
Royal Forest Department
Bangkhen, Bangkok 10900
Thailand

Dr. Peter Kunstadter
East-West Population Institute
1777 East-West Road
Honolulu, Hawaii 96848
USA
(Current address:
Institute for Health Policy Studies
University of California, San Francisco
1326 Third Avenue
San Francisco, Calif. 94143
USA)

Professor James R. Mainoya
Dean, Faculty of Science
University of Dar es Salaam
Dar es Salaam
Tanzania

Professor Walther Manshard
Programme Director
Development Studies Division
The United Nations University
Toho Seimei Building
15–1, Shibuya 2-chome
Shibuya-ku, Tokyo 150
Japan;
Geographisches Institut
Universität Freiburg
Werderring 4
D-7800 Freiburg i. Br.
Federal Republic of Germany

Dr. Ida Bagus Mantra*
Gadjah Mada University
Bulaksumur, Yogyakarta
Indonesia

---

* Invited participant; paper submitted in absentia.

Professor Akira Miyawaki
Department of Vegetation Science
Institute of Environmental Science and Technology
Yokohama National University
Yokohama
Japan

Mr. Sombhan Panateuk
Director
Sriracha Regional Forest Office
Sriracha, Cholburi
Thailand

Dr. Somsak Priebprom
Faculty of Forestry
Kasetsart University
Bangkhen, Bangkok 10900
Thailand

Miss Salakchit Pubiam
Faculty of Forestry
Kasetsart University
Bangkhen, Bangkok 10900
Thailand

Mrs. Kanok-on Reoleung
Faculty of Forestry
Kasetsart University
Bangkhen, Bangkok 10900
Thailand

Miss Kamlai Rirkpipul
Faculty of Forestry
Kasetsart University
Bangkhen, Bangkok 10900
Thailand

Professor Sanga Sabhasri
Permanent Secretary
Ministry of Science, Technology, and Energy
Yothi Road
Bangkok 10400
Thailand

Dr. Puckprink Sangdee
Department of Pharmacy
Chiang Mai University
Chiang Mai 50002
Thailand

Dr. Anant Saraya
Brackish Water Division
Department of Fisheries
Bangkok
Thailand

Mr. A. T. Mahinda Silva
Consultant
Marga Institute
61, Isipathana Mawatha
Colombo 5
Sri Lanka

Dr. Samuel Snedaker
Division of Marine Affairs
Rosenstiel School of Marine and Atmospheric
  Science
University of Miami
4600 Rickenbacker Causeway
Miami, Fla. 33149
USA

Miss Prapasri Thanasukarn
Director
Research Project and Co-ordination Division
National Research Council
Bangkhen, Bangkok 10900
Thailand

Observers

Mr. Prasarn Bamroongrasd
Chief
Mangrove Forest Management Section
Sriracha Regional Forest Office
Sriracha, Cholburi
Thailand

Dr. Twee Hoonchong
Director
Bangsaen Institute of Marine Science
Srinakharinwirot University
Bangsaen, Cholburi 20131
Thailand

Mr. Wichai Suwanpuckdee
Forest Officer
Songkhla Regional Forest Office
Royal Forest Department
Songkhla
Thailand

Mr. Kasem Wongcharoen
Assistant Provincial Forest Officer
Cholburi Provincial Forest Office
Cholburi
Thailand